ARTIFICIAL INTELLIGENCE

Front cover image: © Lidiia/Shutterstock

Published by arrangement with UniPress Ltd, London, by
the MIT Press

AI: Modern Magic or Dangerous Future?

© 2023 UniPress Ltd, London

Text © Yorick Wilks

For image copyright information, see p.224

Designed by Wayne Blades

ISBN 978-0-262-54545-7

Library of Congress Control Number: 2022936049

Printed and bound in China

The MIT Press

Massachusetts Institute of Technology

Cambridge, Massachusetts 02142

http://mitpress.mit.edu

ARTIFICIAL INTELLIGENCE

MODERN MAGIC OR DANGEROUS FUTURE?

YORICK WILKS

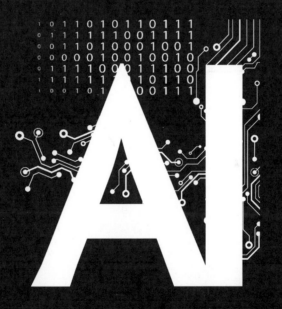

The MIT Press

Cambridge, Massachusetts

CONTENTS

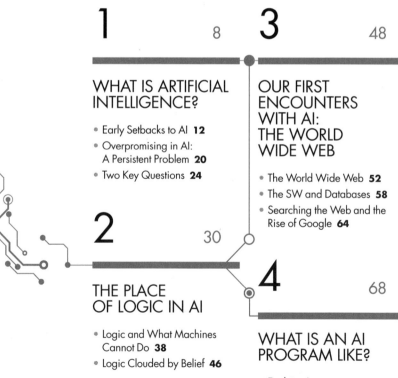

A TIMELINE OF AI

1950
Alan Turing proposes what we now call the Turing Test. (See page 18)

1960
Ray Solomonoff lays the foundations of a mathematical theory of AI.

1974
Ted Shortliffe's PhD dissertation on the MYCIN program demonstrates a practical rule-based approach to medical diagnoses, even in the presence of uncertainty. (See page 43)

1979
The Stanford Car becomes the first computer-controlled, autonomous vehicle when it successfully circumnavigates the Stanford AI Lab.

BKG, a backgammon program written by Hans Berliner at Carnegie Mellon University defeats the reigning world champion.

1952
Arthur Samuel writes the first game-playing program for checkers. (See page 111)

1963
Edward Feigenbaum and Julian Feldman publish *Computers and Thought*.

1965
Joseph Weizenbaum builds ELIZA.

1981
Danny Hillis designs the connection machine, which utilizes Parallel computing to bring new power to AI, and to computation in general.

1951
The first working AI programs are written for the Ferranti Mark 1 machine.

1961
Unimation's industrial robot Unimate works on a General Motors automobile assembly line.

1958
John McCarthy creates the LISP. (See page 78)

1972
The PROLOG programming language is developed by Alain Colmerauer at Marseille. (See page 80)

1985

The autonomous drawing program, AARON, created by Harold Cohen, is demonstrated at the AAAI National Conference.

1986

The team of Ernst Dickmanns at Bundeswehr University of Munich builds the first robot cars, which can drive at up to 55 mph (89 kph) on empty streets.

1998

Tim Berners-Lee publishes his Semantic Web Roadmap paper. (See page 52)

1999

Sony introduces the AIBO, a domestic robot similar to a Furby. It is one of the first artificially intelligent autonomous "pets."

2011

IBM's WATSON computer defeats the champions of the television game show *Jeopardy!*

Apple's Siri (2011), Google's Google Now (2012), and Microsoft's Cortana (2014) are smartphone apps that use natural language to answer questions, make recommendations, and perform actions.

2018

Alibaba language processing AI outscores top humans at a Stanford University reading and comprehension test.

1994

With passengers on board, the twin robot cars VaMP and VITA-2 of Ernst Dickmanns and Daimler-Benz drive more than 620 miles (1,000 km) on a Paris three-lane highway in heavy traffic at speeds up to 80 mph (130 kph).

The Deep Blue chess machine (IBM) defeats the world chess champion, Garry Kasparov.

2002

iRobot's Roomba® autonomously vacuums the floor while navigating and avoiding obstacles. (See page 141)

2020

Microsoft introduces its Turing Natural Language Generation (T-NLG), the "largest language model ever published at 17 billion parameters."

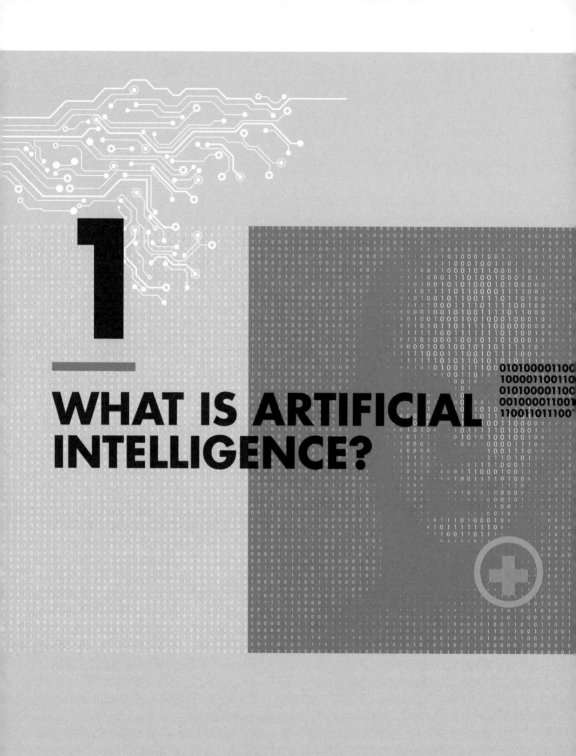

1

WHAT IS ARTIFICIAL INTELLIGENCE?

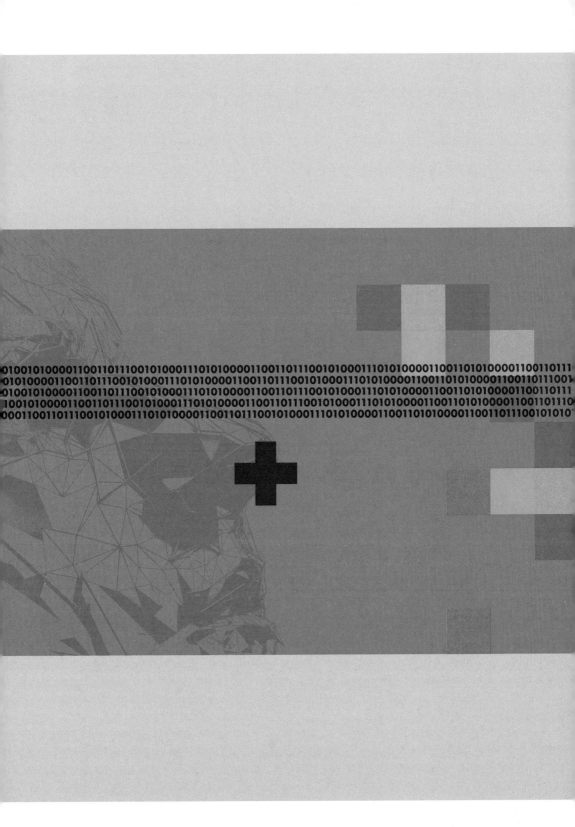

The Austrian philosopher Ludwig Wittgenstein wrote that a philosophy book could be written consisting entirely of jokes. In that spirit, an AI book could perhaps be written now consisting entirely of snippets from the daily news making claims about breakthroughs and discoveries. But something would be missing: how can we tell which of them are true, which really correspond to programs or are just science fiction fantasies? In this book, my aim is to describe the essence of AI now but also to give an account of where it came from over a long period, and to speculate about where it may be headed.

← Perseverance, the Mars Rover launched by NASA in July 2020.

In 2019, there were reports of a Belarusian woman who had programmed her dead fiancé's texts into a "neural network" with which she could now talk posthumously. The same idea appeared in "Be Right Back," a 2013 episode from the *Black Mirror* TV series, and in an article I wrote for *Prospect* magazine in 2010 called "Death and the Internet." The same meme comes back all the time and is appealing, but did the Belarusian programmer really do anything serious? No one knows now, but the need to distinguish research—pouring from companies and laboratories—from speculation, fantasy, and fiction has never been greater, and I try to sort them out in this book.

The term "artificial intelligence" was coined in 1955 by American computer scientist John McCarthy—one of the handful of AI pioneers whose reputation still grows—in a proposal for a 1956 conference at Dartmouth College, in New Hampshire. But doubts about the phrase have grown since then, and English cryptographer and computer scientist Donald Michie's earlier version, "machine intelligence," is making a comeback. The term was notably revived when the journal *Nature Machine Intelligence* began publication in 2019, conferring a badge of scientific respectability on a sometimes-dubious field, where the word "artificial" has come to have overtones of trickery.

McCarthy said firmly that AI should be chiefly about getting computers to do things most humans do easily and without thinking, such as seeing and talking, driving, and manipulating objects, as well as planning our everyday lives. It should not, he said, be primarily about things that only a few people do well, such as playing chess or the Japanese board game Go, or doing long division in their heads very fast, as calculators do. But Michie thought chess was a key capacity of the human mind and that it should be at the core of AI. And the public triumphs of AI, such as beating Garry Kasparov—the then world champion—at chess in 1997, and more recently by playing world championship Go, have been taken as huge advances by those keen to show the inexorable advance of AI. But I shall take McCarthy's version as the working definition of AI for this book.

There can, then, be disputes about exactly what AI covers, as we shall see. I shall take a wide view in this book and try to give a quick and painless introduction to its history, achievements, and aims—immediate and ultimate.

EARLY SETBACKS TO AI

It is important to see how long AI has been gestating, slowly but surely, and to appreciate that the ride has been bumpy, with major setbacks. In 1972, the American philosopher Hubert Dreyfus published a book called *What Computers Can't Do*, where he called AI a kind of *alchemy* (forgetting for a moment that alchemy—an early form of chemistry positing that metals could be transformed into each other—has actually turned out to be true in modern times with the discovery of nuclear transmutation!). Dreyfus's central point was that humans learned as they grew up, and only creatures that did so could really understand as we do; that is to say, be true AI. The philosopher's theories were rejected at the time by AI researchers, but his criticisms had an effect on their work and understanding of what they were doing—he helped rejuvenate interest in machine learning as central to the AI project.

In 1973, Sir James Lighthill, a distinguished control engineer, was asked by the British government to examine AI's prospects.

He produced a damning report, which shut down research support in the UK for AI for many years, though some work continued under other names, such as "intelligent knowledge-based systems." Lighthill's arguments about what counted as AI were almost all misconceived, as became clear years later. He himself had worked on automated landing systems for aircraft, a great technical success, and which we could easily now consider to be AI under the kind of definition given earlier—the activity of simulating uniquely human activities and skills.

Lighthill considered that trying to model human psychology with computers was possible, but not AI's self-imposed task of 8just simulating human performances that required skill and knowledge. He was plainly wrong—the existence of car-building robots, automated cars, and effective machine translation on the Web, as well as many AI achievements we now take for granted, all show that. Although a philosopher and an engineer respectively, Dreyfus and Lighthill

A nineteenth-century illustration of the
Babbage Difference Engine.

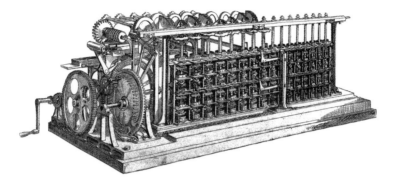

By the early 1900s, the London Brick
Company had automated molding clay
into bricks, after which they were
transported manually to the kiln.

had something in common. Both saw that the AI project meant that computers had to have knowledge of the world to function. But for them, knowledge could not simply be poured into a machine as if from a hopper. AI researchers also recognized this need, yet believed such knowledge could be coded for a machine, though they disagreed about how. (We shall revisit this topic—of knowledge and its representation—many times.) Dreyfus thought you had to grow up and learn as we do to get such knowledge, but Lighthill intuited a form of something that AI researchers would describe as the "frame problem" and he thought it insoluble.

The frame problem, put most simply, is that parts of the world around us "update" themselves all the time depending on what kind of entity they are. If you turn a switch on, it stays on until you turn it off, but if it rains now, it may not be raining in an hour's time. At some point, the rain will stop. We all know this, but how is a computer to know the difference—that one kind of fact that is true now will stay true, but another will not be true some hours from now? As we grow, humans learn how the various bits of the world are, but can a computer know all that we know and function as we do? At a key point in the 1982 science fiction movie *Blade Runner*, a synthetic person, otherwise hard to tell apart from a "natural" human, is exposed as such because he does not know that when a tortoise is turned over, it cannot right itself (see opposite).

The frame problem is serious and cannot be definitively solved, only dealt with by degrees. There have been many attempts, in AI and in computing generally, to prove that certain things *cannot be done*. Yet, in almost all cases these proofs turn out to be, not false, but useless because solutions can be engineered to get round the proofs and allow AI to proceed on its way. According to legend, Galileo, when before the Inquisition and told firmly that the Earth could not possibly move, muttered under his breath the words "*Eppur si muove*"—"and yet it moves"! Marvin Minsky at MIT, one of the great AI pioneers, once said that it is hard to spot AI progress sometimes, but when you come back ten years later you are always astonished at how far it has moved.

The ghosts haunting AI over the years, telling its researchers what they cannot do, recall the "proofs" given that machine translation (MT) was impossible.

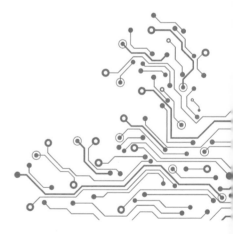

BLADE RUNNER
1982

Blade Runner, a cult sci-fi movie from 1983, presents audiences with a futuristic world of near-perfect humanlike robots called "replicants" with objectives far removed from those of their human counterparts. It explores how these near-human robots might be detected by the humans around them, and looks at the ethics concerning their treatment. In the movie, the process of identifying possible replicants involves posing a single, seemingly simple question: "If you turn a turtle over, what happens?" It is assumed that replicants will not know that turtles cannot right themselves, although this is something that many modern people may not know either!

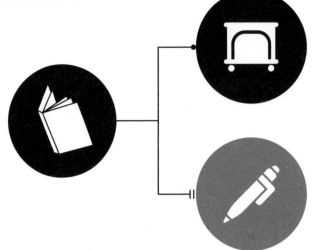

←

In English especially, many or most words are ambiguous: here, "pen" can mean writing pen or playpen.

MT is another computer simulation of a very human skill that we could now consider a form of AI. In 1960, the Israeli philosopher Yehoshua Bar-Hillel argued that MT was impossible, because to translate language the system would have to have an enormous amount of world knowledge. His famous example was *The book was in the pen*, where he argued that a computer would have to know a book could fit into a playpen but not into a writing pen, if it was to get the right sense of the word "pen," and so translate the sentence out of English into some language where those were quite different words. This is an almost exact correspondence with the frame argument mounted against AI. Again, the everyday availability now of free MT of reasonable quality, from sources such as Google Translate,

shows how wrong Bar-Hillel was, though he was very influential at the time and widely believed.

I led a team in 1997, in New York, that entered and won the Loebner competition, at which an annual prize was awarded for the best computer conversationalist. The competition was set up so that journalists had short conversations at computer consoles, behind some of which were real people answering back, while behind others were the competitors—computer conversational programs. The job of the jury of journalists was to rate them all on "humanness" and so decide on the best program of the year. How our team did this—and the kinds of tricks we used to fool the judges—included such things as making deliberate spelling mistakes to seem human and making sure the computer responses came up slowly on the screen, as if being typed by a person, and not instantaneously as if read from stored data. Here was the start of one of the conversations in New York that convinced "Judge 4" that our *Converse* program was a human:

CONVERSE:
Hi, my name is
Catherine,
what's yours?

JUDGE 4:
My name is Elliott.

CONVERSE:
Well, it's nice to meet
you, Elliott.
Did you see that story
on CNN last night
about the lesbian
couple who came out
at a White House party
on Sunday?

JUDGE 4:
Yes, I did. I think it
may all be a publicity
stunt for Ellen
[DeGeneres].

That output is now more than twenty years old, and there has not been a great deal of advance since then in the performance of such "chatbots." This annual circus derived from English mathematician Alan Turing's thoughts on intelligent machines in 1950 (see the box on page 18), and his original test of how we might *know* a machine was thinking. His paper "Computing Machinery and Intelligence" laid the groundwork for seventy years of discussion of the philosophical question:

"Can a machine think?"

When we talk to others we never ask if they are machines or not. It does not make for a good conversation if you ask your friends that kind of thing. Nor did Turing think that we should ask the same question of machines. That issue had to be implicit. Yet now, in competitions such as the Loebner, the question "Are you a computer?" has come out into the open, and contestant machines are programmed to deal with it and give witty replies, as do current commercial virtual assistant systems such as Alexa and Siri. But that question is no longer any real test of anything except ingenuity.

My reason for mentioning the Loebner competition is the curious feature that the level of plausibility of the winning systems has not increased much over the past twenty years. Systems that win do not usually enter again, since they have nothing left to prove. So new ones enter and win but do not seem any more fluent or convincing than those of a decade before. This is a corrective to the popular view that AI is advancing all the time and at a great rate. As we shall see, some parts are, but some are quite static, and we need to ask why, and whether the answer lies in part in the optimistic promises researchers constantly make to the public and those who fund them.

THE TURING TEST

Turing modeled his "test" on a Victorian parlor game in which a contestant would ask questions, via folded notes passed from another room, with the aim of establishing whether the person answering the questions was a man or a woman. In the game Turing proposed, sex detection was still the aim, and if no one noticed when a computer was substituted, then the computer had in some sense won; it had been seen as a person. The crucial point here is that the game was about men versus women—no one knew a computer might be playing. The irony, when we consider how his test has been adapted to events such as the Loebner competition, is that Turing was not trying to say that computers did or ever would think. Instead, he was trying to shut down what he saw as useless philosophical discussion and present a practical test such that, if a machine passed it, we could just agree that it thought and so could stop arguing fruitlessly about the issue.

ALAN TURING
1912–1954

Alan Turing is the modern father of AI. He defined what could and could not be done by possible computations, and was the first person to investigate how we would determine whether a conversational machine might be considered "human." He proposed what is now called the "Turing Test," which has come to mean that if you cannot tell if your conversational partner is a machine, you might as well allow that it is intelligent. He thought there was no philosophical way of settling the matter and a simple practical test was better.

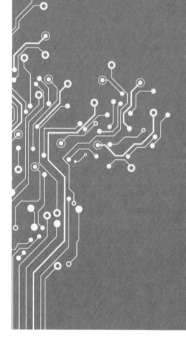

OVERPROMISING IN AI: A PERSISTENT PROBLEM

Coming to grips with this issue is important because it is becoming harder to separate what AI has done from what it promises—and from what the media thinks it promises. There are also science fiction worlds that are close to ours but hard to distinguish from reality. In the 2013 movie *Her*, Scarlett Johansson's voice was given to a "universal AI girlfriend" who seemed able to keep up close conversational relationships with millions of men worldwide. Since speaking and listening technologies such as Alexa, which are sold all over the world, listen to their owners even when they are not attending to them and then report their conversations back centrally, one can ask whether the public knows that Alexa exists but that the Johansson fiction does not? The makers of sex robots are working hard to bring something like "Her" into existence. We shall need to be clear in what follows about what is known to work, what isn't working—yet—and what may never work, no matter how hard we try.

Sorting these things out is made harder not only by company promises, made to sell products, but by researchers who must constantly overpromise what they can do to win public research grants—a problem in the field since World War II. Already in the 1940s, when the capacity of the biggest computer was a millionth that of a smartphone, the papers were full of claims of "giant brains," reasoning and thinking and being just about able to predict the weather for months to come. As early as 1946, *The Evening Bulletin* in Philadelphia wrote of the technology at the University of Philadelphia that a "30-ton electronic brain at U of P thinks faster than Einstein." It was all nonsense, of course, but there was real progress, too.

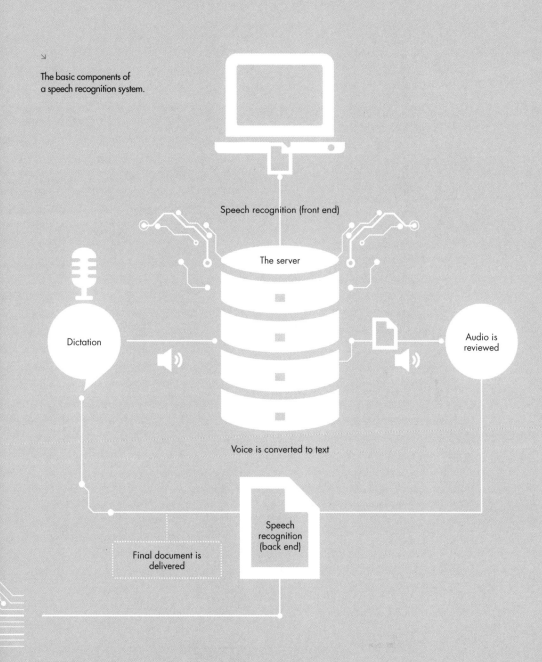

The basic components of a speech recognition system.

Speech recognition (front end)

The server

Dictation

Audio is reviewed

Voice is converted to text

Final document is delivered

Speech recognition (back end)

Someone said recently that the most striking thing about today, to anyone who came here directly from the 1980s, would be that you could have something in your pocket that knew virtually everything there was to know. Think just how astonishing that is, let alone that the device also makes calls. We marvel now at driverless automobiles, but computers have been landing planes without problems for nearly forty years. One of my tasks here will be to convey which parts of AI are moving rapidly and which seem a little becalmed.

GPS

OLDER THAN YOU THINK

The history of AI is important, because although it now seems everywhere, at least according to newspapers and other media, and is pressing upon every human skill, it has been around for a long time and has lapped up around us very slowly. Here is a dramatic example—the traffic sign below was at the end of the driveway of the Stanford University AI laboratory when I was there in the early 1970s.

This vehicle, which was rarely seen, consisted of four bicycle wheels with a wooden tray on top holding a radio antenna, a camera, and a computer box. It could be steered by radio but sometimes ran itself around the driveway, steered by the onboard computer. The device was far more significant than its absurd appearance. It was the beginning of the US-government-funded "Moon Lander," later the "Mars Lander," project, which was set up because it was known that vehicles on either celestial body would have to be autonomous, or self-driven. They would be too far away from Earth to be radio-controlled; they might fall down a crevasse in the time it took for a radio signal to reach them. That primitive Stanford vehicle, which ran fifty years ago, is the father of all the autonomous vehicles now being tested on our highways and covering millions of miles a year.

AUTOMATED CARS

Automated vehicles have come a long way since Stanford University's robotic vehicle of the early 1970s (see the box opposite). Modern iterations make use of an array of sensors and inputs when in motion.

LIDAR
(Light Detection and Ranging)

Camera

RADAR
(Radio Detection and Ranging)

Ultrasonic sensors

Central computer, EMAPS
(Enhanced Mapping and Positioning System)

TWO KEY QUESTIONS

Two key questions will run, and hopefully be answered, throughout this book.

1

Should AI be simply using machines to imitate the performances humans give or should it be trying to do those things in the same way we do them, assuming we could know how our brains and bodies work?

These two things can be quite different, with the first often thought of as engineering and the second as a kind of operational psychology—explaining ourselves to ourselves by using computers. For example, some programs designed to determine the grammatical structure of English sentences process them from right to left—that is, backwards. They imitate an English reader's performance but by methods we can be fairly certain differ from theirs, and so could not be models of an English reader's functioning.

It has long been a truism in AI thinking that, since the Wright brothers made the first human flight, airplanes fly but not with anything like the mechanism of flapping wings that birds use; and this example has been used to stress the difference between modeling the mechanism of evolution—of birds, in this case—and really doing engineering. But more recently, the metaphor has reversed, because it is now possible to build drones that fly as birds do, and moreover to model in them the change of wing shape that enables many birdlike maneuvers impossible for conventional planes.

MODELING OR ENGINEERING?

Institutions around the world are investing in the study of avian biomechanics—in particular wing morphology— in an attempt to enhance aircraft maneuverability. The research has other applications, too. For example, it informs ongoing efforts to reduce the noise aircraft make by introducing subtle changes to their body shape.

2

→

The three main branches of AI, from
the electrical processes of cybernetics,
through to a logic-driven intermediate
phase, to the statistical base of
machine learning.

Should AI be based on building representations inside computers of how the world is, or should it just be manipulating numbers to imitate our behavior?

The current fashion in AI is for the second approach, called machine learning (ML), or even deep learning (DL), and many of the current news items in the media are about applications of this approach, such as recent successes in diagnosing diseases or instances where a computer beat the world's best Go player. Those are approaches based on numbers and statistics. But up until about 1990, the core AI approach used a form of logic to build representations—what I sometimes call "classical" AI: structures representing things such as the layout of scenes or rooms. This is still how applications such as GPS work, by internally examining structures of city streets to find the best route to drive. Such systems are not making statistical guesses about how the streets of a city such as London are connected.

This is an ongoing argument in AI research. When John McCarthy was offered statistical explanations back in the 1970s, he would say: "But where do all these numbers come from?" At the time, no one could tell him, but now they can, as we shall see in later chapters. One way of looking at the issue of those who want to represent things logically, versus those who think statistics a better guide to doing AI, is to remember how AI emerged. It was once bound up with a subject called cybernetics, a word now rarely used in the English-speaking world, though it is still used in Russia and in parts of western Europe. Cybernetics was about reaching the goals of AI not with digital computers but with what were then called analog computers, based not on logics but on continuous electrical processes, such as levels of current. The science of cybernetics produced things such as "smart" home thermostats and mechanical tortoises that could learn to plug themselves into wall sockets—they did not have

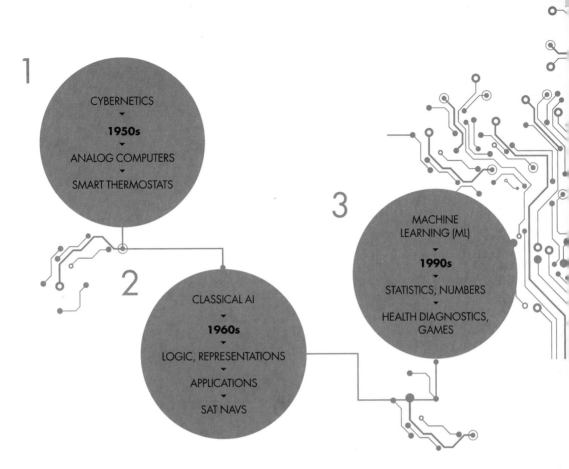

1

CYBERNETICS
▾
1950s
▾
ANALOG COMPUTERS
▾
SMART THERMOSTATS

3

MACHINE
LEARNING (ML)
▾
1990s
▾
STATISTICS, NUMBERS
▾
HEALTH DIAGNOSTICS,
GAMES

2

CLASSICAL AI
▾
1960s
▾
LOGIC, REPRESENTATIONS
▾
APPLICATIONS
▾
SAT NAVS

representations in them at all. With the rise of classical, logic-based AI in the 1960s, in which reasoning was a central idea, cybernetics faded away as a separate subject. But the history of AI still has it jostling for space with other close disciplines, such as control engineering (which pioneered planes that land automatically), pattern recognition (which introduced forms of machine vision), and statistical information retrieval.

There is nothing odd historically about subjects jostling against each other, disappearing in some cases (such as phrenology), or emerging from each other, as psychology and much of science did from philosophy in the 1800s. It is a little like prehistoric times, when different tribes

of humans—Neanderthals, Denisovans, *Homo sapiens*—coexisted, competed, and interbred before one won out conclusively.

The case of information retrieval (IR) is important because of its link to the World Wide Web—the system of documents, images, and video we all have access to via our phones and computers. Google still dominates all searches on the Web, and the company's founders, Sergey Brin and Larry Page, conceived their searching algorithm in the Stanford AI laboratory as part of PhDs they never finished. Although it came from within an AI laboratory, Brin and Page's search method was also directly within classic IR, with a subtle twist I shall describe later.

The relevance of this to our big question is that IR, like cybernetics, does not deal in representations in a way that makes logic central, as "classical" AI did.

Karen Spärck Jones was a University of Cambridge scientist who developed one of the basic tools for searching the Web, and once argued that AI has much to learn from IR. Her main target was classical AI researchers, whom she saw as obsessed with content representations, when they should—according to her—have been making use of the statistical methods available in IR. Her arguments are very like those deployed by older cyberneticians, and more recently by those who think machine learning is central to AI. Her questions regarding AI resolve to this crucial one: how can we capture the content of language except with its own words, or other words we use to explain them? Or, to put it another way, how could there be other representations of what language expresses that are not themselves language? This was a question that obsessed Ludwig Wittgenstein in the 1940s, and he seems to have believed language could not be represented by anything outside itself or be compressed into some logical coding. Here is a brief quotation from Spärck Jones in the 1990s that gives the flavor of her case that classical AI is simply wrong when it thinks that computers can reason with logical representations (what she calls the "knowledge base") rather than by "counting words" (another way of describing doing statistics with texts):

> **"The AI claim in its strongest form means that the knowledge base completely replaces the text base of the documents."**

"Knowledge base" here means some logical structure a machine then uses to reason with, rather than the "text base"— that is, the original words themselves. This issue, of what it is that computers use as their basic representation of the world about which they reason, is still not settled.

WILL AI ALWAYS BE IN DIGITAL COMPUTERS, OR COULD IT BE IN BODIES?

A further question touched on toward the end of the book is about whether the basis of AI should be in digital computers at all, as it has been since cybernetics disappeared in the 1970s, or whether we shall reach AI not by copying how humans do things in computers but by merging computation with the biological—with real human or animal body tissues.

For some, and this approach is more popular in Japan than in the West, this implies building up organic tissue-like structures that can perform, an approach that could be parodied as "doing Frankenstein properly." The alternative, more popular concept among American thinkers and entrepreneurs such as Elon Musk, is called "transhumanism," the view that we could improve humans as they are now with artificial add-ons so that such beings gradually become a form of AI—and possibly become immortal.

All these possibilities are full of religious and ethical overtones—of the creation stories in the Bible, of early artificial creatures such as the Golem of Prague, and of ancient quests for immortality. I shall touch on serious questions such as this in Chapter 10.

Most eye-catching developments in recent AI, from medicine to playing Go, and translating on the Internet, are based on ideas closer to Spärck Jones and IR than to the logics and "knowledge" on which AI was based for its first fifty years.

The next chapters will describe the basic areas of artificial intelligence, including its relationship to the craft of computer programming, and we shall start with asking how important logic is to AI. McCarthy and others believed that AI was about making computer models of logical reasoning in machines and humans—an idea of the primacy of logic in thinking that goes back to the late 1600s and Gottfried Leibniz, the first man to say such things. I shall discuss the scope of machine logic and its decline and fall with the realization that people do not seem to use logic much in everyday reasoning, or even statistics.

2

THE PLACE
OF LOGIC IN AI

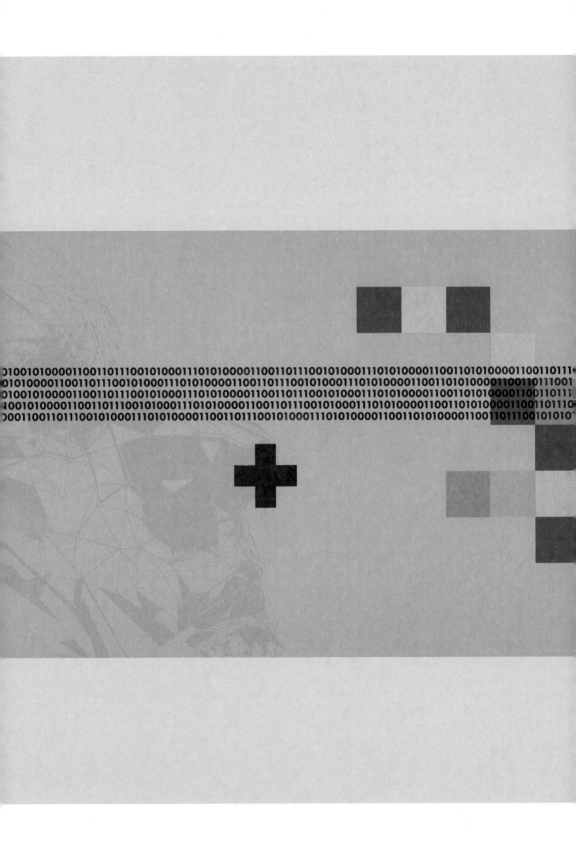

I used to give talks with titles such as "Keeping logic in its place," which now sounds very presumptuous. I did that in reaction to the standard assumption in AI that the core of AI was *reasoning* done by computers, since that was also how human beings functioned. That assumption was almost universal throughout the last quarter of the twentieth century among most AI researchers and developers. Then, AI was seen as the realization of a dream going back to Aristotle, who thought humans were defined by their rationality, and most importantly to Leibniz, the seventeenth-century German philosopher.

Leibniz is most celebrated for inventing the calculus—albeit at the same moment as Isaac Newton—but he was also obsessed by the idea that if humans could only calculate together (the word he used for reasoning) then all problems, social and political, could be solved. He invented a logic of symbols—going beyond Aristotle, who worked with words to reason—and an artificial language in which you could express thoughts and reason about them. In a famous passage, he wrote that if missionaries could only translate the Gospel into this language, heathens everywhere "could no more doubt its truths than the theorems of Euclid." An ambitious project! Leibniz is a crucial step on the path to classical AI—the idea of a language of representation, not like the ordinary languages we speak, but one that could

express what we want to say in some other, more exact way. As we saw in the previous chapter, it was the assumption that one could use a language of representation that Karen Spärck Jones claimed was the crucial fallacy of classical AI. Leibniz had no computers, of course, but he was aware of clockwork machines that could imitate humans; he thought reasoning was, in some important sense, logical and mechanical. In this spirit, later twentieth-century logicians such as Bertrand Russell created much more complex logics of symbols, which foreshadowed programming languages and the representations of AI.

GOTTFRIED WILHELM LEIBNIZ
1646–1716

Gottfried Wilhelm Leibniz was a seventeenth-century German philosopher, logician, and mathematician, whom Bertrand Russell credited as being probably the cleverest man who had ever lived. Simultaneously with Newton he invented differential and integral calculus.

Leibniz's importance for AI is that he was the first person to argue that reasoning is essentially a mechanical process, an insight that was the driving force of much early AI. He was also the first to explicitly explore the notion of infinity and to consider the existence of other logically possible worlds, separate from this one. He argued that God had set the universe determinately in motion, so that everything that happens does so necessarily—by preprogramming, as it were. Leibniz argued that this was the best of the possible worlds God could have created, even though it contained evil.

THEOREM PROVING

Pioneering AI researchers such as John McCarthy also believed that even our simplest thought processes—not just chess and puzzles but writing and planning our day—must rest on a system of logic running in our brains. Given that, the task of AI had to be to capture that reasoning in computers. There were three difficulties with this approach, usually known as "theorem proving," since mechanical reasoning came down to knowing whether a particular sentence/theorem could be proved from another set of sentences or not.

1 Much psychology research has shown that people do not in fact reason logically in everyday life unless trained to do so—through, for example, school exercises.

2 Research in logic often resulted in proofs that "you can't do that," which some claimed to show that logic was not up to the job of doing complete reasoning. The most famous example was Gödel's theorem, which showed that not all true sentences can be proved true, even though people can see they are true.

3 The practical results of theorem-proving research over decades was not impressive in practical terms—it gave us few useful AI systems that worked.

Another recent blow to belief in the role of logic in human life has been the "behavioral economics" pioneered by the Israeli-American psychologist and economist Daniel Kahneman, who showed conclusively that the standard economic model of humans as rational deciders of their own best interests was wildly false and that normal human behavior was systematically irrational. A typical example concerned decisions on where to live in America. Kahneman showed that Californians and Midwesterners were equally happy on some standard scale, but *both* believed that Californians were happier and were prone to act on such a belief by moving house—against all the evidence.

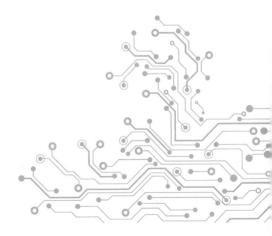

THE WASON TESTS

As an illustration of Point 1, we can look at some ingenious experiments by the English psychologist Peter Wason in the 1960s, which are further explained on page 36. The usual interpretation of these experiments is that people are not good at using logic in everyday problems without training. But if you change to a familiar situation, such as age and drinking, then they know what to do because they understand concrete situations but not logic itself.

Another of Wason's demonstrations involved the sentence:

NO BRAIN INJURY IS TOO TRIVIAL TO BE IGNORED.

He carried this about on a card and let people read it for themselves—he thought that if he read it aloud it would impose an interpretation on it. People were then asked to say whether it meant "Treat all brain injuries" or "Ignore all brain injuries." Nearly all of them voted for the first, though in reality it means the second, as can be seen if you reverse its sense by substituting:

TREATED

for

IGNORED

You can then see that it does indeed mean the first, so it must have originally meant the second, strange though that seems, as the sentence is simple and contains no difficult or ambiguous words. What it does have, however, is a succession of powerful logical words such as

NO, TOO,

and

IGNORED

(meaning *don't do it!*).

Wason's explanation was that our mental processor finds this succession of logical moves too complicated to follow, so it simply gives up and plumps for the socially acceptable explanation, which is to treat injuries, not ignore them. Again, the message is that the human mind cannot work very well with logic, but goes directly for what it is used to, or for what it thinks acceptable.

PETER WASON'S
LOGIC EXPERIMENTS

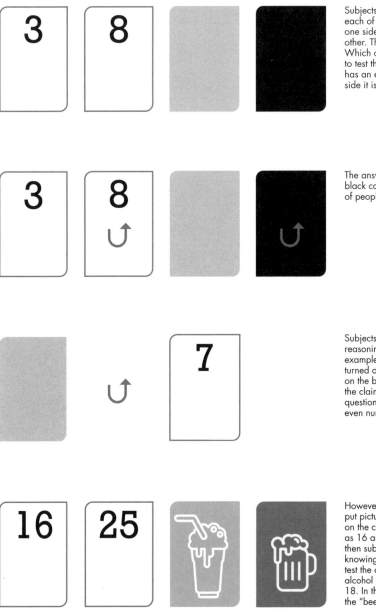

Subjects are shown four cards, each of which has a color on one side and a number on the other. They are then asked: Which cards must you turn over to test the claim that if a card has an even number on one side it is gray on the other?

The answer is the 8 card and the black card, but only 10 percent of people typically get that right.

Subjects fail to do the logical reasoning to show that, for example, if the gray card when turned over has an odd number on the back that does not show the claim wrong, because the question is about cards with even numbers on one side.

However, when the researchers put pictures of beer and soda on the cards, and ages such as 16 and 25 as the numbers, then subjects are much better at knowing which cards to turn to test the claim if you are drinking alcohol then you must be at least 18. In this case you need to turn the "beer" card and the 16 card.

PETER WASON
1924–2003

Wason was a twentieth-century British psychologist much concerned with how humans represent knowledge and how they understand. He thought they did this with practical examples from everyday life and not by means of logic, as many in AI believed. After serving in the British Army during World War II, Wason graduated in English at Oxford University, later reading psychology at University College London.

LOGIC AND WHAT MACHINES CANNOT DO

I have touched on the way AI has always been dogged by arguments about what computers cannot do, even in principle, and some arguments concern logic and what can be proved. Gödel's theorem is the most famous of the arguments and refers to two proofs Austrian-born logician Kurt Gödel gave in the 1930s about limits to the power of logic. The proofs actually concern arithmetic but are taken to apply more generally. In the first theorem, Gödel showed that, for any logical system beyond the simplest, there would always be true statements it could not prove. This was a huge blow to those who thought logic was *complete*, meaning that all true things could be proved.

Gödel had a second theorem that showed that a logical system could not prove itself *consistent*, which is to say that it did not contain sentences that contradicted each other. The relevance of this to AI was thought to be that if logic and reasoning by computer were to be the core of AI, but you could never know for sure whether a given sentence could be proved true from other sentences, there would be a problem. It needs a bit of technical argument to get to that from the two theorems, but imagine we have a set of sentences and want to prove one of them. For our purposes, let us refer to this sentence as p. Now imagine we added to the set a sentence that was the opposite of p—this can be termed NOTp. In logical terms, our original task of proving p from the original set is almost the same as proving that the second set (which contains both p and NOTp) is inconsistent—that is, it contains a contradiction. But if we cannot prove that the contradiction is there—and Gödel's theorem says we cannot—then we cannot be sure we can prove the original sentence p. All that means is that you cannot be certain of proving everything that is true; it certainly does not mean that proving things with computers cannot work.

Again, we are in the situation where proving that things cannot be done in all cases and everywhere does not mean that those things cannot be done as much as is

"A CONSISTENT FORMAL SYSTEM ABLE TO PRODUCE ARITHMETIC CANNOT PROVE ITSELF CONSISTENT."

↑

A visualization of Gödel's first theorem: a human can recognize the sentence above to be true, whereas a logic machine (the red circle) cannot.

necessary for practical purposes. What held up computer proofs was not this kind of demonstration of impossibility but that the work of sorting through huge numbers of assumptions to find the ones relevant to what you were trying to prove was so time-consuming. But ingenious AI researchers found all kinds of shortcuts for doing just that.

Some philosophers jumped onto Gödel's proofs and tried to show they implied that computers could never have the capacities of humans—because there were things we could see were true but which could not be proved by logic, making full AI impossible. Arguing this, however, ignored the question of how humans knew those things were true in the first place. The argument assumed that computers must use logic to know things were true. But they could, of course, use other ways of finding out truth, such as statistical guesses—just as we do. We know most of what we know without proof and, since we do not know it by magic or occult powers, there must be ways in which we do this. Much of what we "know" is what we were taught (and we believed), yet some of that will certainly be false—so we do not really know it. Why, then, should not machines do the same things we do to find out truths, such as accepting what teachers say?

It is tempting, faced with evidence such as Kahneman's on real human choices, to jump too far the other way and declare that logic is entirely the invention of scholars since Greek antiquity, and that it is all an educational illusion with no place in human life. But that would be quite wrong—modern civilization rests on all kinds of rational processes and has done so since long before computers, right back to the construction of the pyramids and campaign plans of Roman armies. Some things simply have to be thought out and planned in fantastic detail. Imagine building a jet engine from thousands of parts without a clear logical plan of what order to do things in.

There has always been another tradition in philosophy, saying that the power of logic was overstated where everyday life was concerned. David Hume, arguably the greatest British philosopher, wrote in the eighteenth century:

"And if [ideas about facts] are apt, without extreme care, to fall into obscurity and confusion, the inferences are always much shorter in these disquisitions, and the intermediate steps, which lead to the conclusion, much fewer than in the sciences…"

I take Hume to be saying that, outside the sciences, where deduction and logic have a real role, inferences are quite "short," consisting of only a few steps. Moreover, they are informal steps rather than true logical steps, although Hume does not say they are informal because, 200 years ago, he did not have the contrast we now have between human mental processes and formal logic or programming languages. But the kinds of inferences described above that people use to sort out which card to turn over to check a drinking age suggest he had in mind what we would now call informal inferences, which are brief and take place in something very like language itself.

KURT GÖDEL
1906–1978

Kurt Gödel may be the most influential logician of the twentieth century. He produced proofs that no formal system of interest could prove all the truths in its field—and therefore it was in a strict sense incomplete. This was a great blow to those who thought logic could be the foundation of mathematics by proving all its truths. The problem arose as to how one could know those "extra" statements were true, if not by proof. Some argued this showed that humans were therefore superior to any logical machine because they could see the truth of such statements but machines could not. This forgot that computers do not need proof to know things are true and can use probabilities, just as humans do. He spent his last years reexamining the old proofs of God's existence.

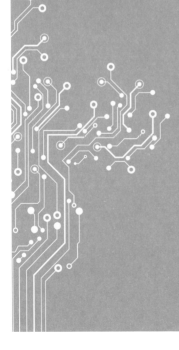

Robot-assisted surgery

$40B

Assistance to nursing

Administrative support

Detecting fraud

Reducing errors of dosage

Connected machines

Identifying participants for clinical trials

$20B

$18B

$17B

$16B

$14B

$13B

Initial diagnosis

Automated image diagnosis

Cybersecurity

$5B

$3B

$2B

ESTIMATED ANNUAL BENEFITS OF
HEALTH AI APPS BY 2025

AI TESTING GROUND

Like the early robot vehicle (see page 22) and the original Google algorithm, the most striking expert systems, with names such as DENDRAL and MYCIN, came from Stanford University—in particular, the laboratory of American computer scientist Edward Feigenbaum. DENDRAL was a system to identify new forms of potential organic compounds that might be worth synthesizing in the laboratory. As the name expert systems suggests, these efforts were not the computer modeling of everyday life skills such as language and vision. Instead, they were based on coding into rules the knowledge of scientific and medical experts, so as to ultimately outperform them— which these systems did in the 1970s and 1980s. MYCIN was the first truly effective computer system for diagnosing infectious diseases. It was also the first AI system to encounter the inevitable and powerful resistance from the medical profession to the challenge it posed, which it has taken nearly forty years to overcome. It is vital to stress that such expert systems were not pure, theoretical theorem-provers at all, but rather worked with shortcuts, usually called *heuristics*— procedures that trimmed and made practical the searches for evidence to prove conclusions. Furthermore, the power of such systems was in the expert knowledge they contained rather than in the logic itself. As Feigenbaum put it: "Intelligent systems derive their power from the knowledge they possess rather than from the specific formalisms and inference schemes they use." This work also gave impetus to a separate psychological discipline called "knowledge elicitation," a method for getting experts to reveal their knowledge so it could be coded— sometimes knowledge they did not even know they had.

←

Ed Feigenbaum, Director of the Computation Center at Stanford University, and his team working on expert systems in 1966.

VICTORY FOR AI

A much publicized AI success from the last decade illustrates informal inferences. In 2011, IBM's WATSON system beat the best human contestants at the American TV game show *Jeopardy!*, in which participants are given an answer and have to guess the corresponding question. So, given the clue:

"Its largest airport is named after a World War II hero and its second after a World War II battle,"

the correct response would be:

"What are Chicago's airports named after?"

WATSON's answer was not based on logic and reasoning, except to a minor degree; it was a combination of IR—a skill we met already, here used to search through the enormous quantity of document data the system held—and what is called natural language processing, or NLP. This is the area of AI that, like machine translation, deals with the manipulation and production of written language, a topic we shall discuss in much more detail in Chapter 5. WATSON would take the clue words, such as "airport" and "World War II hero," and search in classic IR fashion for documents that contained them, and then try to isolate the sentences that might contain

the answer. It would then make quite short inferences, in the style we associated with Hume (see page 40), to show that some string or words it had found did have the clue as its answer. WATSON was highly successful and a triumph, not for logic at all but for IR and NLP.

WATSON received huge publicity and was touted by IBM as its entry point into medical AI—a way of providing doctors with expertise in response to questions put to a huge medical literature. But two important misunderstandings about the WATSON experience should be addressed here:

1   It was not a breakthrough technology at all, but a refinement, over a long period, of state-of-the-art NLP technologies combined with IR—in particular, locating relevant sentence fragments and forming them into a proper response, then showing that it answered a question. This work had been done over decades as part of DARPA-funded projects. DARPA, originally ARPA, is the US Defense Advanced Research Projects Agency, the main historical supporter of AI research. It was also the US government agency that had funded the Internet, and much of US NLP.

2 Although WATSON performed well in the competition in real time, its victory was almost certainly due not only to its processing speed, but to the speed with which it could press the buzzer first. The human response is delayed because of the time taken for the nerve signal from the brain to travel to the finger on the buzzer. WATSON had no brain and no finger and could answer in far fewer milliseconds than any human. That was almost certainly decisive and, in hindsight, WATSON's speed should have been adjusted for fairness.

WATSON was a qualified success and showed the importance of limited uses of logic, in this case combined with complex text processing. Another fact that made possible the later revival of statistical and nonrepresentational machine learning methods in the 1990s was that more thorough logical methods than WATSON produced few products that captured the imagination of either researchers or the public. However, a scaled-down version of the logic program—renamed "expert systems"—did have some limited success, chiefly in science and medicine, as well as in specialized areas such as the use of logic to optimize the distribution of computers across a given floor space.

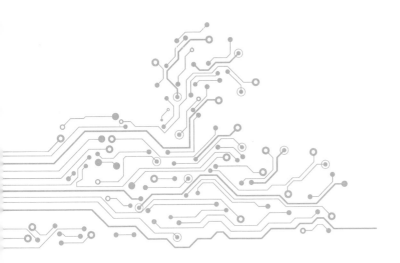

LOGIC CLOUDED BY BELIEF

Processing beliefs to identify different descriptions of the same person.

We should look briefly at another key aspect of logic that has been important in creating AI representations of the world and our knowledge of it. A fundamental principle of logic since Leibniz has been the ability to substitute alternative names for the same thing in a logical form *without changing the meaning or truth of the whole*, whether we call that a formula, an expression, or a sentence. So, if Joe is the name of Jane's father and I write: "Joe is married to Ann," that will be just as true if I substitute the alternative words for "Joe" to get the sentence: "Jane's father is married to Ann"—because "Joe" and "Jane's father" are the same person. A large part of logic is devoted to the problems that arise with the simple move of adding in human belief to create sentences such as: "I believe Joe is married to Ann." The problem arises because that may not be true at the same time as: "I believe Jane's father is married to Ann" is true, although they are the same person. It is simply that I do not happen to know the *name* of Jane's father.

A whole industry has grown up within logic to deal with this issue, called *opacity* (meaning contexts you cannot see through). But what is important to AI, in its desire to model human representations of the world, is that it must have some practical way to deal with this issue of belief if it is to model individual humans conversing with each other. Such individuals will almost certainly have different knowledge states and different, inconsistent, beliefs about the world, such as who is married to whom. It is obvious we can talk to people who do not share all our beliefs—about democracy as much as about family members—and one way of describing that is to say we have models of each other in our minds and consult these so as to communicate. I talk to you based on a model I have of you, and of what you may or may not believe, which may differ from what I believe. A teacher must have this model to teach a student who, by definition, does not yet know or believe the things that the teacher does. Similarly, a doctor will have to talk to a patient who refers to a pain in their

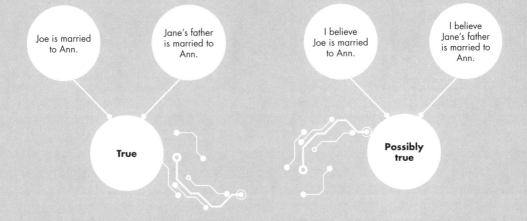

stomach but who points to the wrong place because they have a wrong belief about where their stomach is (as many people do, believing it to be lower than its real position).

In later chapters I shall discuss individuals and their own knowledge and beliefs in relation to the Internet, and I will assume that AI can construct and manipulate representations, or *models*, of these individuals' beliefs to allow computers to talk to them effectively. The notion of a machine *having a belief* will therefore be an important one for AI, though not one I want to use carelessly. I do not want to say an ATM has a belief about how much is in my bank account. I want rather to say it just has data about me. But if a computer could have a model of me in which it could see that I thought I had more in my account than I do, and could therefore correct me gently about that, I might want to say it did have a belief *because it could contrast two points of view of my account and see how they differed.* This idea of machine belief as having alternative points of view is

an important inheritance from logic—one that AI needs—but not one that logicians ever developed much for themselves. AI is not just a handmaid of logicians but has sometimes developed logical ideas further within itself. Yet, as we shall see, logic has been vital to AI in another way—in the construction of special programming languages to construct AI systems. The languages PROLOG and LISP—which I shall discuss in Chapter 4—have been crucial to the development of AI programs and were developed directly from ideas in logic. The AI–logic relation is a complicated and mixed one that can be characterized as being about practicality versus theory, but which is also one of great indebtedness to more than 2,000 years of intense thought.

3

OUR FIRST ENCOUNTERS WITH AI: THE WORLD WIDE WEB

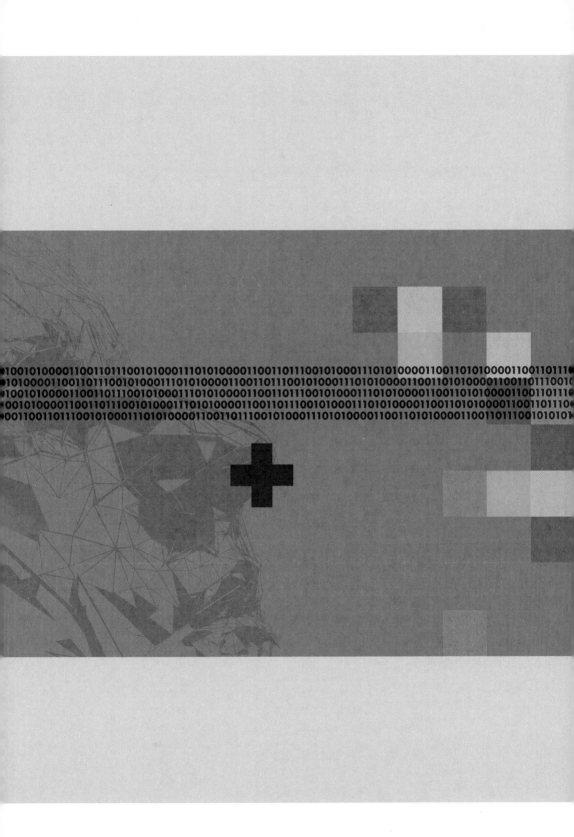

How do people encounter AI in everyday life? First, almost certainly when they use the Internet on their computer or phone to search for information. Then when they sit in a plane landing blind in fog, or when having surgery done by a robot. Within a few years many of us will be in self-driving cars.

The overall purpose of this book is to give some idea of what kinds of AI ideas lie behind these machine abilities. Hollywood movies and the media constantly obscure the line between science fiction AI and the AI that really exists. So, what is the best AI we really have right now? Rather than answer that directly—and list the marvels and promises that fill the media—I would first like to look back at the history of the World Wide Web and ask how we got to where we are, a situation where our phones can reach pretty much all the world's available information from public information sources as well as the 10,000 pictures of food we took but will never look at. I assume that we cannot separate the modern history of AI from that of the World Wide Web, which is its main interface to the public.

The Web is a good example of an unintended consequence—something more common in science and technology than you might guess. It is similar to the many cases of things being designed for one purpose but actually found to work for another. British psychologist Richard Gregory claimed in the 1960s that the brain was an engine designed largely for seeing but which has had its "seeing mechanisms" adapted later to yield our language faculty.

It is not always understood that many of the elements we associate with the Web were available on the earlier forms of the Internet when it was still dominated by its origins as a military-academic communication system set up by DARPA in the US. There was already the ability to shift large files round the Internet from site to site—the protocol was not the current Web protocol (called http, which you see in all Web addresses) but the much older FTP, the "file transfer protocol." I can remember pioneering AI worker Marvin Minsky at MIT sending out the whole of a draft book to everyone on the Internet in 1972, the first time such a thing had been done.

The Internet itself—or the ARPANET as it was originally called—was the brainchild of American computer scientist J.C.R. Licklider and designed for US military requirements, which needed to get data and information seamlessly across a network of US government computers. It was, in computer science terms, a "packet switching network," breaking up information into tiny packets of bits and sending them off across a network of cables or microwaves and then reassembling them into documents at the destination. It was decided early on to include the great AI research laboratories such as Stanford and MIT on the network. The ARPANET thus hugely accelerated the progress of AI research. Soon after email was invented in the late 1960s, it drew the US AI research community into a single whole, though even then, just as now, much of the email traffic was already about personal needs and assignations.

The ARPANET later became the Internet and was made worldwide and "civilianized." But before that, and long before the arrival of Tim Berners-Lee's World Wide Web, which was carried by the Internet, there were facilities such as Usenet. This was a system of peer-to-peer file transfer with no central organization, which supported a huge range of user groups open to anyone, and which sent images as well as text—all interests were catered for. Blogging certainly started there, rather than on the later World Wide Web, and many people set up file sites designed to encourage users to read personal diaries.

THE WORLD WIDE WEB

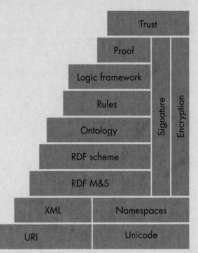

				Trust		
				Proof		
				Logic framework		
				Rules	Signature	Encryption
			Ontology			
			RDF scheme			
			RDF M&S			
	XML		Namespaces			
URI		Unicode				

The arrival of the World Wide Web in 1990 changed everything. English computer scientist Tim Berners-Lee's original vision, conceived at CERN's international physics laboratory in Geneva, Switzerland, was to give every document a physicist might want a name (later called its URL, or Uniform Resource Locator), so that the document could be seen from anywhere in the world, regardless of where it was stored. This stimulated the creation of new ways to search the names of Web servers and their files, none of which were part of the original World Wide Web design itself. Berners-Lee, when he started the Web, just had a list of servers holding documents, until that number became too large to be manageable.

↑

The classic levels of processes on which the Semantic Web rests.

THE SEMANTIC WEB

Earlier I described the World Wide Web as a good example of an unintended consequence. This is because the Web that Berners-Lee gave the world in 1993 was not what he had originally intended—what you might call Berners-Lee's first great idea was that of a Semantic Web (SW). The Web we actually got was serendipitous, like Post-it® Notes, because it did not have the key property that he was looking for in the SW, which was to express the meaning of what it contained. The SW is no longer simply an aspiration but

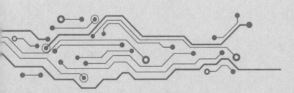

is now a serious research subject on both sides of the Atlantic, with its own conferences and journal. Even though it may not yet exist as a single demonstrable entity in the way the Web itself plainly does, it exists in part and is a topic for research and about which fundamental questions can be asked as to its representations and their meanings.

The World Wide Web is like TV. It has millions of documents and images on it but does not know what is in them, just as a TV does not know what it is showing—it just puts it on the screen. The SW conceived by Berners-Lee was a system of documents and images where the system itself knew what they contained and *could reason about* their content. The SW therefore is much more like AI than the Web we have because it could show intelligence.

In his paper in the *Scientific American* in 2001 that set out his design for the SW, Berners-Lee described making an appointment for an elderly relative, a task that would require planning and access to both the doctor's and relative's diaries and synchronizing them. This kind of planning behavior was at the heart of what I have been calling classical AI. If you look at the SW diagram from the original *Scientific American* paper (see page 52), it shows the SW as a stack of "levels of representation," which just means symbols that express the meaning of other symbols. That is very like what dictionaries do for us—we look up a word and get other words telling us what the word means. In the SW stack, the "lowest" level is the base in the Internet that carries the World Wide Web, and would carry the SW.

The tendency has always been to look at the upper levels of this figure—rules, logic framework, and proof—and it is these that have caused both critics and admirers of the SW to say that it is the classic AI project by another name. But looking at the lower levels, you find "Namespaces" and "XML," which are products of what I have called NLP (natural language processing), obtained by annotations of texts by a range of NLP technologies, a subject I shall discuss in detail in Chapter 5.

Markup languages such as XML grew up in humanities departments as ways of marking, say, that a name in a text was a proper name (perhaps that of an Egyptian pharaoh). This technique was taken over by NLP researchers to mark texts with their grammar, such as marking a particular noun as a noun. That sounds obvious, but later we shall see that this was done so that a program could learn a grammar, if it was shown enough examples of what a noun was—by putting marks into a text, such as <proper noun> after a noun such as "Ptolemy."

It is important to remember that available information for science, business, and everyday life still exists overwhelmingly as text—85 percent of business data is still unstructured data (what "text" is now technically called). The same is true of

A sample paragraph of text alongside its LaTeX type setting.

```
\documentclass{ article} % Starts an article
\usepackage{amsmath} % Imports amsmath
\title{\LaTeX} % Title

\begin{document} % Begins a document
  \maketitle
  \LaTeX{} is a document preparation system for
  the \TeX{} typesetting program. It offers
  programmable desktop publishing features and
  extensive facilities for automating most
  aspects of typesetting and desktop publishing,
  including numbering and cross-referencing,
  tables and figures, page layout,
  bibliographies, and much more. \LaTeX{} was
  originally written in 1984 by Leslie Lamport
  and has become the dominant method for using
  \TeX; few people write in plain \TeX{} anymore.
  The current version is \LaTeXe.

% This is a comment, not shown in final output.
% The following shows typesetting power of LaTeX:
\begin{align}
  E_0 &= mc^2 \\
  E &= \frac{mc^2}{\sqrt{1-\frac{v^2}{c^2}}}
\end{align}
\end{document}
```

L^{A}T$_E$X

L^{A}T$_E$X is a document preparation system for the T$_E$X typesetting program. It offers programmable desktop publishing features and extensive facilities for automating most aspects of typesetting and desktop publishing, including numbering and cross-referencing, tables and figures, page layout, bibliographies, and much more. L^{A}T$_E$X was originally written in 1984 by Leslie Lamport and has become the dominant method for using T$_E$X; few people write in plain T$_E$X anymore. The current version is L^{A}T$_E$X2ε.

$$E_0 = mc^2 \tag{1}$$

$$E = \frac{mc^2}{\sqrt{1-\frac{v^2}{c^2}}} \tag{2}$$

information on the Web, though the proportion of it that is text (as opposed to diagrams, photos, videos, and tables) is falling. And you might ask, how could the Web become the SW, except by information being extracted from natural text and stored in some other form along with it? NLP in the 1990s created technologies that could store, within the text, indications of what its parts meant and what their structure was. The inspiration for this did not come only from the humanities, but also from printing technology—an early sideline in AI laboratories, and computer science more generally, in the 1970s was printing documents on the computers then available. This was done using systems for inserting marks into a document's text, to indicate how that which followed the mark should be printed, such as putting it in italics or in a larger font.

The most sophisticated of these systems was a printing language called TeX (later, LaTeX), invented by American computer scientist Donald Knuth at Stanford, originally to manage the look of his own work on the page, but subsequently becoming the main tool mathematicians used to lay out and print their formulas in their papers. At about the same time, humanities scholars developed The Text Encoding Initiative (TEI) for the scholarly markup of text, which then developed into Standard Generalized Markup Language (SGML), and then into HyperText Markup Language (HTML), which, with its successor Extensible Markup Language

(XML), has become the standard for encoding pages on the Web. Later, special markup languages were developed for time (TimeML), speech (VoiceML), and a plethora of other special areas. They all share the idea captured by those in-text printing commands: that you insert special marks into text that are not part of the text itself, but which tell you what to do with the text, and how to interpret it, or what it means. These "marked-up" pages, created by the page designer, underlie every Web page you see—they are the bones beneath its easy-on-the-eye flesh.

From this standpoint, the SW can be seen as a conversion from the World Wide Web of texts by means of an annotation process of increasing depth, one that projects notions of meaning up the SW diagram (see page 52) from the bottom. In 1953 the English philosopher of science R.B. Braithwaite wrote an influential book, *Scientific Explanation*, on how scientific theories obtain the semantic interpretation of "high-level" abstract entities (such as neutrinos or bosons in physics) from low-level data in observations. He named the process one of semantic ascent up a hierarchically ordered scientific theory. The view of the SW under discussion here, which sees NLP as providing its foundational processes that provide the interpretation to be given to the "higher-level" concepts in the Web, what is usually called its ontology, bears a striking resemblance to that view of scientific theories in general.

TOO MUCH INFORMATION

Sometimes consequences can turn bad, or at least dubious.
The French MINITEL system, a precursor of the Internet, was
designed in the late 1970s to give access to French phone
directory information on screens in homes. It became, however,
largely a tool for the sex industry, with Paris Métro trains
in the 1980s full of ads with erotic MINITEL call signatures.
It is now a familiar observation that the French suffered the
disadvantage of early innovators, and the wide reach of
MINITEL made the later advance of the Internet slower in
France because the country already had a popular system
with some of the Internet's features.

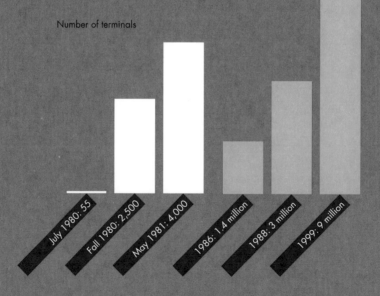

Number of terminals

July 1980: 55   Fall 1980: 2,500   May 1981: 4,000   1986: 1.4 million   1988: 3 million   1999: 9 million

USAGE FIGURES OF MINITEL IN FRANCE

↑
A French MINITEL terminal, originally for finding phone numbers, and later a limited local version of the World Wide Web, but which in the end held back French progress.

THE SW AND DATABASES

Berners-Lee's views of what the real SW was to be—the one he really wanted and not just the Web we now have—developed with the availability of very large databases into what is often called the era of "big data." Some now see databases as the core of the SW: databases, the meanings of whose features are kept constant and trustworthy by a cadre of guardians of their integrity, where "features"

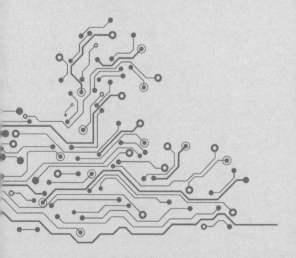

just means the indicators, always in English words, of what the database stores, such as BIRTHDATE. This integrity is important because words change their meanings with time—for example, we still "dial numbers" when we make phone calls, even though that no longer involves the finger actions it did a few decades ago. So not even number-associated concepts are safe from time. Computer scientists seem to believe that, unlike ordinary language, the features of data can be preserved from changes of meaning, so that the content or interpretation of data cannot change. It is just this aspect of AI that Karen Spärck Jones was arguing against when I quoted her in the previous chapter (see page 28). She saw preserving meaning not as a compliment, but rather as a hopeless and impossible romantic dream.

Spärck Jones's view of AI owes something to a theory the American philosopher Hilary Putnam put forward in the 1970s. Putnam believed scientists were guardians of meaning, since only they know the true

chemical nature of, say, the metal molybdenum and how it differs from the superficially similar aluminum. So, said Putnam, only scientists really knew what those words mean. For him it was also essential that the scientists did not allow the criteria of meaning (for being, say, aluminum as opposed to being molybdenum) to leak out to the public, lest they then became subject to change. Thus, for Putnam, only scientists knew the distinguishing criteria for separating water and deuterium dioxide (heavy water), which seem the same to most of the population but are not. I personally do not think such a separation (of popular meaning from that known to expert guardians) makes sense, quite aside from how profoundly undemocratic it is.

The heart of the issue about content can also be put as follows:

We have the feeling that we locate things in our minds by content, that we search by what something means. The SW represents an attempt to realize that feeling in the world of information search.

Spärck Jones argued that whenever that desire is realized, we later find that some statistical method, which makes no reference to meaning, can do the job better. That is the "guardian of content" issue at the SW's core, and it remains unresolved—with head and heart at odds with each other.

←

Do only scientists know the meanings of aluminum and molybdenum since only they know the science that distinguishes them? To everyone else, they seem exactly the same.

BERNERS-LEE'S ORIGINAL WEB IDEA

My argument that there was an element of serendipity about the creation of the World Wide Web differs from the conventional view, which is that Berners-Lee first set up a "web" of documents (the World Wide Web) and then shifted his interests toward what we now call the SW. I believe this is not correct and that he always intended something like the SW. Here is Berners-Lee's own account, from 1998, of what he initially did in Geneva:

"… in 1989, while working at the European Particle Physics Laboratory, I proposed that a global hypertext space be created in which any network-accessible information could be referred to by a single 'Universal Document Identifier' … I wrote in 1990 a program called 'Worldwide Web,' a point-and-click hypertext editor which ran on the 'NeXT' machine. This, together with the first Web server, I released to the High Energy Physics community at first, and to the hypertext and NeXT communities in the summer of 1991."

Berners-Lee's historic innovation in the World Wide Web—and it is important to remember, too, the other key developments, such as American computer scientist Marc Andreessen's Mosaic browser, and hypertext itself, an idea of American philosopher of technology Ted Nelson—was that of a document's *unique identifier*. But identifier of what? In his recollections, quoted above, Berners-Lee uses the phrase "Unique Document Identifier" as a simple name—one that has disappeared from the lexicon and that does not appear in Wikipedia, for example.

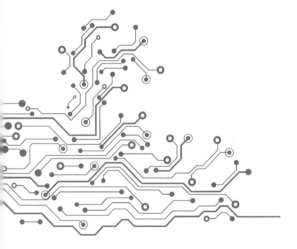

HILARY PUTNAM
1926–2016

Hilary Putnam was one of the most eminent twentieth-century American philosophers of science and logic. He is best remembered for questioning the essential properties of things, such as a cat being essentially an animal, and the connection of this to two kinds of statement, one open to evidence and the other, called analytic, which is true by definition, such as "cats are animals." Putnam asked us to consider how we would react if it were discovered that all cats are actually robots controlled from Mars. Would we, he asked, stop calling them cats because cats are, by definition, animals? He suggested we would just go on calling them cats, in which case the sentence "cats are animals" is not true by definition, but just a kind of fact.

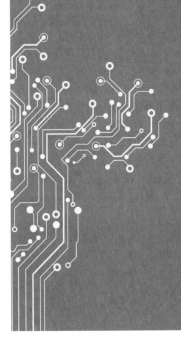

It was soon replaced with the familiar URL (Uniform Resource Locator), which is normally taken as the unique name of a document. All this leads naturally to the Web as we know it, as a web of documents. However, in his recollections, Berners-Lee then goes on to say:

> **"The great need for information about information, to help us categorize, sort, pay for, own information, is driving the design of languages for the Web designed for processing by machines, rather than people.** *The web of human-readable documents is being merged with a web of machine-understandable data [my emphasis].* **The potential of the mixture of humans and machines working together and communicating through the Web could be immense."**

This is precisely the spirit of the original SW document Berners-Lee wrote later with other authors in the *Scientific American*, but even in 1989 he added a diagram in which you can see the lines from documents pointing to objects in the real world. That is not just a web of documents, such as the World Wide Web, but also of *things*—data, or data standing for things, a matter much closer to his heart, as a database specialist, than language in documents. That, I believe, is the evidence for his real view of the SW, even early on— the web of data; data for machines rather than human readers.

So URLs became the piece of Web technology that got into everyday speech, being the name for a document that is, say, a book and that will pick out or point to that book's content, stored on the Web. That name does not pick out a unique real book, an object weighing a couple of pounds—in the way that "Tim Berners-Lee" *does* pick out a unique object, namely him—because there may be thousands of copies of any book, and only the RFID tag (derived from radio-frequency identification) that the bookseller puts in it to stop shoplifting identifies an individual copy. These ambiguities over the names of general classes of things, as opposed to particular individuals, are familiar in the history of philosophy and meaning, and it is illusory to think that the SW has actually solved two-millennia-old problems about names and how they connect to the world of things. But it has given us a vast working example of how a system of names and things can function practically, which is another example of refuting the "you can't do that" arguments we encountered earlier.

INTERPRETATION OF BERNERS-LEE'S DIAGRAM

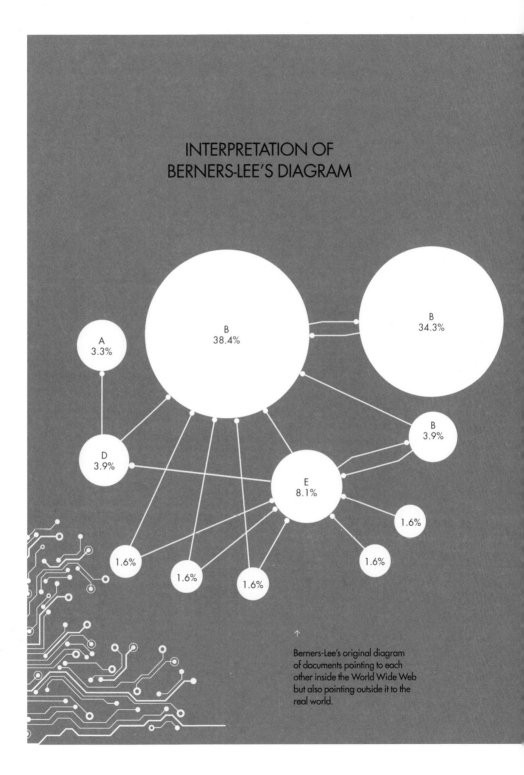

↑

Berners-Lee's original diagram
of documents pointing to each
other inside the World Wide Web
but also pointing outside it to the
real world.

SEARCHING THE WEB AND THE RISE OF GOOGLE

I have concentrated so far on the prehistory of the Web, on how we got here, but the crucial element in most people's encounters with the Internet is *searching*, or how to find things. That task was transformed by two Stanford graduate students, Sergey Brin and Larry Page, whose approach to the search question (see page 27) was so successful that they never finished their doctoral theses on it. Instead, they gave themselves up to Google, the company they founded with their big idea. That company is now one of the most valuable in the world and provides more than 90 percent of all Web searches outside China.

Nothing about Google's search system is wholly original—others were playing with the ranking idea at the same time—but Page and Brin commercialized it and it took over the world of search. At one point, it looked as if the pair had discovered a working definition of what was true on the Web—a "Google View of Truth" as you might be tempted to call it—that in all the welter of information and falsehood with which the Web soon filled up, what was most pointed to would be the most true or reliable. That view was more alluring when, for the first decade or so of its life, the Web was largely in the hands of scientists for whom truth was both important and familiar. As we all now know, things began to change, and ingenious people found ways of gaming the Google PageRank method. There were many cases of people agreeing to point to each other's work and links being for sale or giving precedence in ranking to advertisers. There was always a risk in the original design that a site became popular because all the many links to it were bad ones, such as complaints about service! All these possibilities came and went and were combatted in various ways by Google and its rivals. In Chapter 10, I shall turn to the pressing problem that has arisen as the Web has come to be seen as a source of "fake news" as much as of truth and information.

I have said little here about how search engines such as Google's do their search, beyond assuming that for decades there have been standard IR methods for searching documents based on their so-called key words. Karen Spärck Jones, with others, developed the principal method based on the ratio of significant or rare words in documents to the words most common in all documents in that language. These methods are usually called "bag of words" since they take no account of the order of words in documents, so they have no connection at all to understanding the texts. Although I do not know exactly how Google searches documents these days, because it is a commercial secret, I know it certainly does go beyond bags of words to find exact quotes.

THE SECRET OF GOOGLE SEARCH

Brin and Page's notion of search came straight from IR, even though it was done in an AI laboratory, and their original idea was not about search itself but about ranking—in what order of importance the list of results should be presented to a questioner. They patented their idea as "PageRank," punning on one of their names, and it went as follows: rank the highest—that is, seen first, at the top of the list—those documents that are themselves most linked to by other documents, particularly when those other documents are also much linked to. To put it crudely, a document is more important if many other documents point at it, particularly if those documents in turn are much pointed at. This intelligent network aspect of PageRank is what makes Google search into AI. Google search also needs the notion of a hyperlink, an idea of Ted Nelson's, incorporated into Berners-Lee's World Wide Web, namely that a document can contain within itself a URL—the name of another document to which it points.

URL

http://www...

1ST

2ND

3RD

←

The foundation of Google was the page-ranking algorithm that made a document more important the more other documents pointed to it on the Web.

An old favorite example of this is searching Google for both "models of measurements" and "measurements of models," which, if only bags of words were used, would produce the same list of documents. In fact, the lists returned have almost nothing in common, and the first list is entirely scientific, which is the clearest evidence that search engines now have gone way beyond IR in their use of NLP, which I shall discuss further in Chapter 5.

The origins of the Web are in Berners-Lee's databases for physicists and unconnected to AI. But the sophisticated search capacities of the Web developed from within AI and it is, I believe, moving inevitably if slowly toward some form of SW, one where representations of knowledge and meaning will be central. And those have always been at the heart of AI. The gigantic computer structures of all medical and scientific knowledge on the Web are now indispensable for researchers, and their structure and interpretation have come from decades of work in the mainline AI tradition. Google has created what it calls its "Knowledge Graph," which powers many of its applications and is essentially an application of the SW idea.

Another major point about the SW project and its relevance to AI is the philosophical issue of *grounding*—how our symbols in language or in AI programs link to real objects in the world that *ground their meaning*. Philosopher Ludwig Wittgenstein's treatment of the problem assumed that it was wrong just to think of this as being about how the word "apple" links to a real apple. He suggested it is rather a question of how the vast web of language, with all its links from word to word, connects to the whole world. The Web is just a network of links from words to documents and pictures, and we can ask now if anything in it has any real meaning. What we call URLs end up pointing to yet more symbols and more documents; so how can we ever escape from the world of symbols, to some firm ground? Perhaps the success of the Web, and even more that of the nascent SW, is that we do not have to have that escape for the whole to be useful and to give us the answers we need. In some sense, the Web's very functionality shows that the ancient philosophical problem of attaching words to things does not stop our understanding of the world of symbols. It has always been posed in terms of words such as "apple," but our real lives are full of quite different words, such as "Germany," "courage," "love," "lies," "procrastination," and "Hamlet," which point at nothing at all that is concrete in the world—and yet we cope.

We shall now take a step back and explore the notion of a computer program upon which all AI ultimately rests. Then, we shall move on to particular areas of AI.

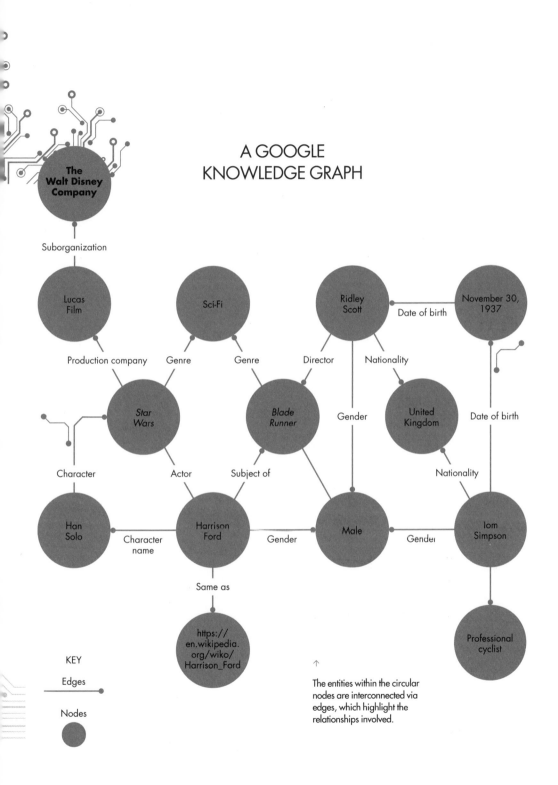

A GOOGLE KNOWLEDGE GRAPH

The Walt Disney Company

Suborganization

Lucas Film

Sci-Fi

Ridley Scott

November 30, 1937

Date of birth

Production company

Genre

Genre

Director

Nationality

Star Wars

Blade Runner

United Kingdom

Date of birth

Character

Actor

Subject of

Gender

Nationality

Han Solo

Harrison Ford

Male

Tom Simpson

Character name

Gender

Gender

Same as

https://en.wikipedia.org/wiko/Harrison_Ford

Professional cyclist

KEY

Edges

Nodes

↑

The entities within the circular nodes are interconnected via edges, which highlight the relationships involved.

4

WHAT IS AN AI PROGRAM LIKE?

01010000110
10000110011
01010000110
00100001100
110011011100

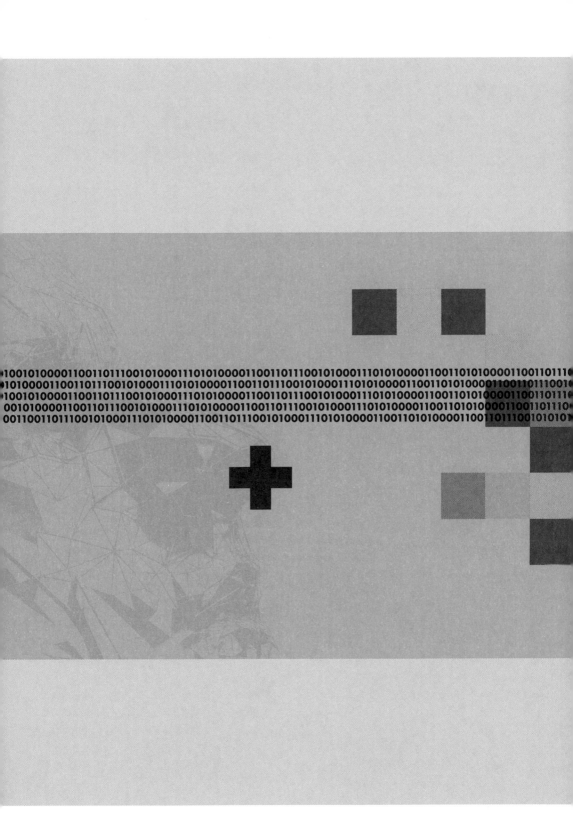

AI practice rests on computer programs. If we forget early cybernetics for now, as well as recent explorations in Japan of computers linked to biological materials, that means programs written for digital computers by human beings.

A digital computer is based on vast banks of switches, which can be in one of two positions. Many other devices can be used as computers: English mathematician Alan Turing got his earliest insights into AI from considering a "Turing Machine" consisting of a long reel of tape divided into squares, along with a stamp that could print and remove numbers in the squares as the tape ran backward and forward. He showed that such a primitive device could do any possible computation. As a graduate student in 1962, I wrote my first program—one to give grammatical structure to sentences—on a Hollerith Machine, which worked by sorting punched cards into slots. It had been designed to manage elections, and it took some ingenuity to turn it into a computer by repeated sorting of the packs of IBM punched cards. All such peculiar ingenuities were typical of the early days of machines, but we can now assume all computers are safely digital.

We are also in an age when governments fret about "coding" and wonder if the whole population should be able to practice it as a useful skill—a bit like cooking or algebra at school. They wonder if, somehow, the population would be more likely to take up a computer programming career or have better relationships with their phones or PCs if they knew in more detail what a computer program was. I see no reason at all to believe the latter any more than we need to know all about car engines to drive. But if we are to set out a basis for AI in this book, it will be a good idea to give some understanding of what a program is.

The word "algorithm" is often thrown about in the discussion of information technology as if it were the same as a program, or software in general. Originally, that word meant something precise and mathematical, connected to what could be proved, but nowadays its standard use is for a piece of program or software that carries out some specific task, using a method that could be mathematical, but need not be. Theorists puzzle as to whether, say, two algorithms doing the same task are equivalent, or would take the same time, or use the same number of steps. Software and program are more general terms and simply mean any piece of code that causes a computer to do something, written in

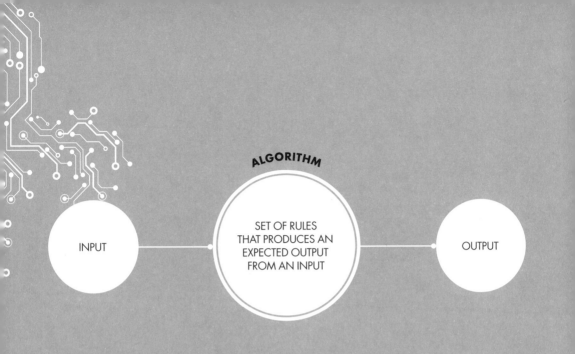

ALGORITHM

INPUT

SET OF RULES
THAT PRODUCES AN
EXPECTED OUTPUT
FROM AN INPUT

OUTPUT

↑
Algorithms employ a step-
by-step method to arrive at
mathematical solutions.

a computer language. A big program can have millions of steps and can be very hard to understand, even for those who wrote it, so programs have *documentation* attached to them, in a language such as English, to explain which bits do which job and why. AI researchers have developed special languages for AI, the best-known ones being LISP and PROLOG, and their style has influenced the form of AI theories since they are often expressed using those languages.

So, a digital computer consists of very large numbers of switches, which can be in one of two positions, usually thought of as

representing the numbers 1 and 0. Collections of such switches can express long strings of 1s and 0s, usually called *binary* numbers (see pages 72–73). All numbers can be expressed in this way, even though the strings tend to be rather long. This is an alternative to how we normally do arithmetic, in "hundreds, tens, and units" with columns that add up to ten. The binary system uses two columns (hence *binary*), where you "carry" a number leftward when you add up to two, not ten.

Binary columns

Binary values added together

STANDARD NUMBERS TO BINARY

128	128	64	32	16	8	4	2	1	
1								1	1
2							1	0	2
3							1	1	2+1
4						1	0	0	4
5						1	0	1	4+1
6						1	1	0	4+2
7						1	1	1	4+2+1
8					1	0	0	0	8
9					1	0	0	1	8+1
10					1	0	1	0	8+2
15					1	1	1	1	8+4+2+1
20				1	0	1	0	0	16+4
50			1	1	0	0	1	0	32+16+2
100		1	1	0	0	1	0	0	64+32+4
150	1	0	0	1	0	1	1	0	128+16+4+2

BINARY NUMBERS

To understand binary numbers, you need to understand the system. To take an example, 1001 is the binary equivalent of the decimal number 9. Each column, going leftward, is a power of two. So, the rightmost 1 means 1, a 1 in the next place left means 2, in the next place left, a 1 means 4, and a 1 in the fourth left column, where we have a 1 in 1001, means 8. There can never be *more than* 1 in a column because it would make a 2 and so get "carried left," just as we do in standard arithmetic when we add 1 to 9. So, 1001 is one 8 plus one 1, which is 9.

←

A table that converts decimal numbers to binary ones.

EVOLVING LANGUAGES

One way of thinking about the binary system is to remember that the Romans—with numbers like MMXXIII (2023)—did not have any columns and so could not easily do arithmetic. This meant that the empire's accounting was done on abacuses, using quite different methods! All that is important here is to understand that strings of 1s and 0s are how a digital machine "knows" anything at all, since the values correspond directly to the settings of its switches. In turn, programs consist of languages that translate whatever they contain into the language of binary numbers so they can run on a machine. Programmers originally had to try to understand such binary strings, which was virtually impossible, but nearly eighty years of work on programming languages has created programming expressions that can be understood by people because they look more like real languages.

The more comprehensible such languages are—which usually means the more they contain terms we can understand and interpret—the "higher" the level of language they are said to be. Languages designed for AI, such as LISP (which I shall describe briefly in a moment), are very high level and must be translated down many times to other languages until they are finally translated into binary numbers to run on the machine. In a slightly satirical way, AI theorists used to ask, "Could English be a programming language?" After all, in armed forces, simple, clear, and consistent commands are given in human language. If I had an intelligent microwave and could shout at it, "Cook this steak medium rare"—and this is now a reality not fantasy—would that mean I was using English as a very high-level programming language? You can answer either way, but most complex procedures cannot easily be reduced to English in the way cooking can, as we shall see when I try to explain the elements of LISP.

The idea of a program—a string of code symbols controlling a complex process— is an utterly modern idea, though first seen

↑
This Roman abacus was one of the first human-made calculating devices.

in the punched cards that drove the looms of French weaver Joseph Marie Jacquard in the nineteenth century, and a little later in the codes written for English mathematician Charles Babbage's incomplete early computer made of brass cogs. But it has now, as a metaphor, infected everything, so that people ask, quite reasonably, "Does the brain have a programming language?" The answer is probably not, though it seems that the genetic code, which drives our heredity, is much more like a language, and one with only four letters. That proves nothing about the brain's function but is at least suggestive that there can be language-like codes "built into" us.

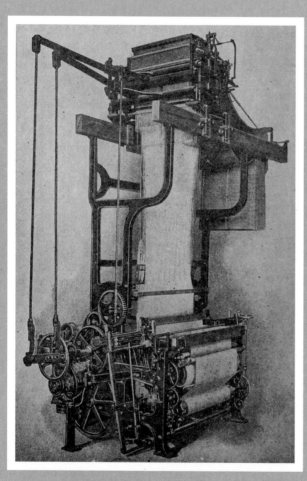

← ↑

A French Jacquard loom (left) where a punched card (above) determined the patterns woven.

Operators of IBM punched card machines (right). IBM pioneered the punched card (below) of coded information, then the size of a dollar bill.

THE LISP LANGUAGE

John McCarthy, whom we have met before, invented LISP. There is no space here to give a proper description of it, but I shall attempt a brief sketch because it was so influential in AI and used so widely to describe its problems and how to solve them.

Most programs consist of commands, to do one thing after another, with commands and loops to repeat actions. McCarthy developed a quite different kind of language, which he called LISP (for *LISt Processing* language), without those commands or loops. Like many modern programming languages, it makes use of functions, where a function is like a black box that takes something *in* (its argument or arguments) and gives something *back* (its value), based on a set of rules. What makes LISP special is that its functions work on lists, with the ability to select items from lists, merge lists, and process them.

The essence of LISP is that you can write a function that uses itself in its own definition. This trick of defining a LISP function in terms of itself is called *recursion*, as opposed to the way that conventional programming languages might work through a series of numbered instructions, which is called *iteration*. You can think of recursion in the following way. Suppose you want to know the length of a list of things. You could say it is *one more than the length of the rest of the list* (the rest after the first item). If you keep cutting down the rest of the list and adding one to your total each time, you will end up with a number that is the length of the list. Yet the italicized phrase above does also define the answer in terms of itself, and that is the key element in the idea of recursion, one that some find mysterious.

Many linguists have claimed that recursion is a fundamental property of the brain and human thought, and it is what distinguishes the grammar of human languages—all of which have it, they argue—from other codes, such as birdcalls and the bits of language that apes can adopt. This is a complex business and has been a key weapon in the fight for and against American linguist Noam Chomsky's theories of language. I shall not go into that here, but it is useful to note the power of this notion of recursion displayed in LISP—

LISP IN ACTION

Imagine a language such as Swahili, where the whole sentence "I do not know" is the single word "Sijui," within which "Si" says it is a "not" (a "not" for me, that is; it would be different for you), "ju" says this is about knowing, and the final "i" says it is *me* talking. If a program needs to discover that this word expresses the idea "know" as "ju," which is buried in the center of the word, it must take this word as a list of letters and find in that list the sequence "ju"— a simple task to express as a LISP process.

JOHN McCARTHY
1927–2011

John McCarthy was a founder of AI research and originated the name itself. He ran the AI Laboratory at Stanford University and was the creator of LISP, which was for decades the main programming language for AI. LISP was based not on commands but the repeated evaluation of mathematical functions. He believed that logic was the underlying reality of all intelligence—animal, machine, and human—a claim now widely disputed. In later life McCarthy tried to create a form of logic that would allow for change and persistence in the world, and so, for example, did not assume that because a light was turned on it would stay on until you were told it had been turned off.

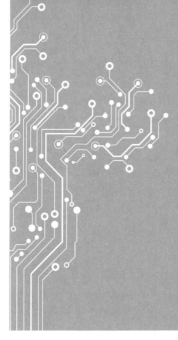

the magic feeling of getting something achieved without having to go through the steps, but just starting out by defining the answer you want and letting it run. This LISP way of thinking—called "top-down" thinking—had a powerful grip on AI for decades when logic and representations were in the ascendant. It is less true now that statistical methods for machine learning have moved to the center of AI—a subject we shall discuss further in Chapter 6.

LISP became so central to the AI of its time that a successful company was founded to build "LISP machines." These were not conventional digital computers but had LISP lists built into their hardware to make such programs run faster. Now that computer hardware is so relatively cheap, and "wasteful" processing is less important, LISP-like languages are having a revival.

Since LISP manipulated lists, and both programs and language sentences are just lists of symbols, it seemed to many that the ideal way to think about processing-language sentences was as lists of symbols—this is what I have been calling Natural Language Processing (NLP) and shall return to at length in the next chapter.

After LISP, PROLOG became another widely used AI programming language, based not on mathematical functions but on logic itself. It too had lists and recursion but consisted of logical expressions and rules, such as:

sibling(X, Y) :- parent_child(Z, X), parent_child(Z, Y)

This rule meant: "If anything is a SIBLING of another, then there is some other thing that is a PARENT of both of them." PROLOG programs consisted again not of commands but of requests to find values that satisfied all the statements of fact and rules the program had as data. Although plainly a form of logic, the language was invented by French computer scientist Alain Colmerauer to do machine translation (MT), and a number of famous MT programs were written in it.

There is a permanent tension in AI between those who want to see AI representations as being more like language or more like logic (aside, that is, from the many others who want them statistical and made up of quantities). These approaches share a common ancestor, since at the time of ancient Greek polymath Aristotle all logic was carried out in a language—Greek—and in a structure called a syllogism. There are many forms of syllogism and the most famous has the form: all men are mortal, Socrates is a man, therefore Socrates is mortal.

HOW LISP WORKS

The LISP language does not have orders like most computer languages, but evaluates mathematical functions as shown.

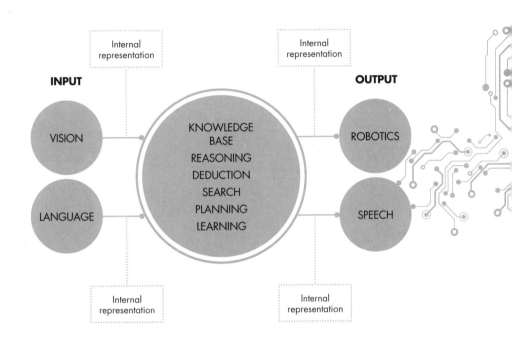

INPUT — Internal representation — Internal representation — OUTPUT

VISION

LANGUAGE

KNOWLEDGE BASE
REASONING
DEDUCTION
SEARCH
PLANNING
LEARNING

ROBOTICS

SPEECH

Internal representation — Internal representation

Logic has reentered our discussion here because a strong strand in modern computing, which has influenced AI as well, has been the desire to make programs reliable in operation by *proving them correct*—that is, proving that a program must deliver the correct answer it was designed to provide. It is understandable that we should want to know not just that a computer program that lands planes in fog always works, but that we can be sure it always will by proving it must do what it says it does. This field of "program proof" was once thought of as part of AI, like automatic programming, but is now part of a separate discipline called "software engineering." This discipline has had substantial success—

sometimes by proving things about programs if they are simple enough, and sometimes by more statistical methods that provide every possible input to a program to see what happens in every case. The method is quite close to machine learning, a subject that has endured within AI.

Programming remains at the very center of AI, since all its applications require it. But, as we shall see later, when discussing machine learning, it is no longer clear who or what has the job of writing those programs. First, though, we need to move from programming languages to AI mechanisms that can understand and produce human languages.

5

TALKING AND UNDERSTANDING: AI SPEECH AND LANGUAGE

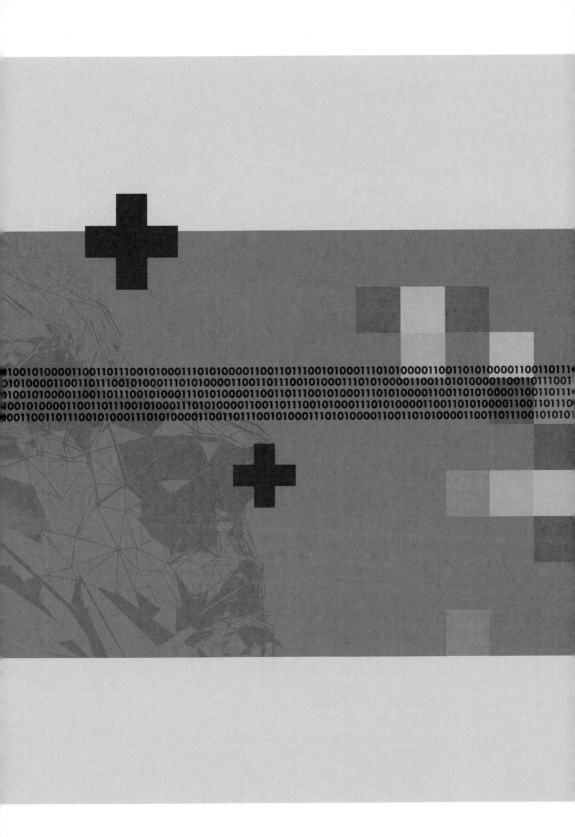

In this chapter I shall introduce programs and machines that talk and understand written language. St Thomas Aquinas famously said that a man could not have an animal as a friend—clearly, he had never had a dog. But not even the most fervent dog lover would claim their animal could talk or listen to anything beyond a handful of simple words. It is only humans who grasp language, and all of us manage speech, so language skills fit firmly into John McCarthy's original goals for AI: those skills we have without thinking or possessing special talent (see page 11).

Despite the current imperfections in speech and language technologies and the clear fact that there is a long way to go before a human level is achieved in AI, the foundations are already in place for entities that talk and understand as we do, up to some reasonable practical level.

Unsurprisingly, the understanding of language and speech by computer has always been a key AI preoccupation, because it is tied closely to our ideas of what mind and intelligence consist of. But their modeling in a computer—getting machines to speak, listen, and understand—has not always been near the center of AI. Their position has depended both on the relation of language to logic, which I have already discussed (see page 28), and on the status at any given time of statistical methods in AI, which tend to discount the unique roles of both language and logic in favor of some notion of statistically expressible information. The recent rise in importance of statistical methods has also

tended to emphasize the differences between the modeling of speech and that of language, since decoding the first (into written symbols) has responded well to such methods but understanding the *content* of both rather less so.

A key element in all such discussion has been the role of machine translation (MT): the task of translating one language into another by computer. This enterprise is as old as AI itself and was not originally seen as part of it. Reasonably good-quality MT is now available free on the Web in dozens of languages and can certainly be counted as an AI success. Those who doubt this should remember how poor much human translation is.

Any theory of language understanding should have implications for improving MT, and so it remains a standard task for evaluating any such theory, even though many would hold that MT does not *always* require a deep level of understanding of what is being translated. Human simultaneous translators

RULE-BASED MACHINE TRANSLATION
SYSTEM WORKING FLOW

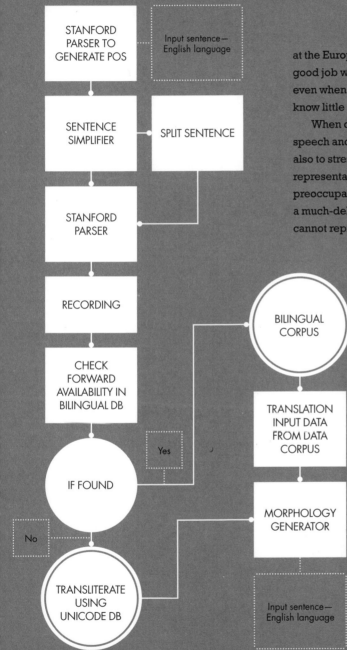

STANFORD PARSER TO GENERATE POS

Input sentence— English language

SENTENCE SIMPLIFIER

SPLIT SENTENCE

STANFORD PARSER

RECORDING

BILINGUAL CORPUS

CHECK FORWARD AVAILABILITY IN BILINGUAL DB

TRANSLATION INPUT DATA FROM DATA CORPUS

Yes

IF FOUND

MORPHOLOGY GENERATOR

No

TRANSLITERATE USING UNICODE DB

Input sentence— English language

at the European Commission seem to do a good job with whatever is being discussed, even when it is a technical subject that they know little about.

When discussing the relation of MT to speech and language research, it is important also to stress that the role of knowledge representation—what we might call the key preoccupation of classical AI—in MT remains a much-debated issue; statistical methods cannot represent it and claim not to need it.

←
The flow of tasks in an early machine translation system based on linguistic rules.

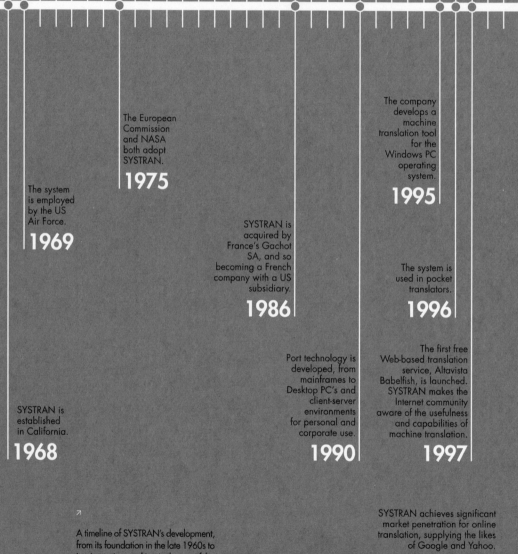

The company develops a machine translation tool for the Windows PC operating system.

1995

The European Commission and NASA both adopt SYSTRAN.

1975

The system is employed by the US Air Force.

1969

SYSTRAN is acquired by France's Gachot SA, and so becoming a French company with a US subsidiary.

1986

The system is used in pocket translators.

1996

Port technology is developed, from mainframes to Desktop PC's and client-server environments for personal and corporate use.

1990

The first free Web-based translation service, Altavista Babelfish, is launched. SYSTRAN makes the Internet community aware of the usefulness and capabilities of machine translation.

1997

SYSTRAN is established in California.

1968

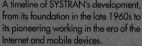

A timeline of SYSTRAN's development, from its foundation in the late 1960s to its pioneering working in the era of the Internet and mobile devices.

SYSTRAN achieves significant market penetration for online translation, supplying the likes of Google and Yahoo.

2002

LOOSE TRANSLATION

SYSTRAN is an American MT system that evolved from the 1950s to become, by the 1980s, the world's most-used system. Originally funded by US defense agencies to translate scientific Russian in the Cold War, it had become an effective tool and was bought by the European Commission to do rough drafts between major languages. It had been reprogrammed several times to make it more adaptable, but it was still a system resting on humans editing failed translations and then storing large numbers of example strings in the source, or starting, language—more than 100,000 in the case of Russian. This was significant because it had always been known that if you could store all the sentences of a language and their translations, MT would be easy. The problem, of course, was that there are an infinite number of sentences in a language; however SYSTRAN's 100,000 Russian strings were a step in the right direction. SYSTRAN was the target that all later MT systems set out to beat if they were to gain attention. It ended up commercialized as Yahoo's Babel Fish service on the Web and remains a robust dinosaur in the MT universe.

The company develops software for the cell phone market.

2009

The company develops a model for self-learning machine translation.

2011

SYSTRAN launches SYSTRAN.io, an API Platform for Easy Development of Multilingual Applications.

2016

Following acquisition by CSLI, SYSTRAN SA forms part of the SYSTRAN International Group.

2014

SYSTRAN expands its on-site translation services for corporate clients.

2015

THE INFLUENCE OF LINGUISTICS

AT THIS EARLY POINT IN THE HISTORY OF NATURAL LANGUAGE PROCESSING (NLP)—THE 1970s AND 80s—THERE WERE FOUR KINDS OF INFLUENCE ON THE DEVELOPMENT OF THE TECHNOLOGY, WHICH WERE MORE OR LESS INDEPENDENT:

Theoretical linguistics, particularly the work of Noam Chomsky, the discipline's most celebrated practitioner, had a substantial influence on the early development of MT, especially his belief that language was basically a syntactic phenomenon: that grammar was central, and considerations of meaning and communication were secondary. Attempts were made to build computer analyzers of syntactic structure based on Chomsky's theories, with a view to later producing machine translation. The best known was a large project at IBM in New York in the 1960s, but all attempts were basically failures, because they produced far too many syntactic structures for any sentence analyzed and there was no way of deciding which was the right one.

1

Theoretical linguistics, as noted above, which had little lasting effect.

2

An AI logic-based approach, which tried to attach logical structures to sentences to show their meaning. Like theoretical linguistics, this produced little of value at any large scale. The most intellectually successful part of this movement lay in the attempts by researchers, such as American computer scientist James Allen, to model what are called "speech acts." This follows a notion from philosophy that what we say is to be understood in terms of what we want to achieve—as when someone might say, "It's cold in here," because they want you to shut the window. These insights led to complex logical models of the goals and intentions of speakers to enable an understanding of input.

3

A movement in AI based on representations of meaning content, made up from items structured into a formal language, but which was not a form of logic or derived from linguistics. The movement was associated with researchers such as American AI theorist Roger Schank and myself and was an idea going back to Leibniz's plan for a universal language of reasoning that was not quite logic. It created systems that formed the basis of some meaning-based MT systems, particularly in Japan.

4

Early research on networks intended to mimic the structure of the brain, called "neural networks." One version was called connectionism, which did not create large-scale systems but was influential in helping revive notions of machine learning that I shall discuss in the next chapter.

THE RISE OF "SHALLOW" METHODS IN NLP

The first extremely basic NLP programs used simple methods to create dialogue with machines, of which ELIZA in the late 1960s is by far the best known, though its contemporary PARRY was much the better performer. ELIZA, created by German-American computer scientist Joseph Weizenbaum, was a simple program that mimicked the kind of psychiatrist who never tells you anything but always asks you about what you had just said, as in: "Tell me more about your mother." The program impressed people at the time and its creator even claimed it was dangerous! But ELIZA fooled no one into thinking it really understood anything that was said to it.

PARRY, developed in the 1970s by a psychiatrist, Kenneth Colby, at Stanford, mimicked a paranoid patient in a military hospital. It conducted thousands of hours of dialogue on the early version of the Internet, including the following short sample:

Have you been hospitalized before?

THIS IS THE FIRST TIME

How long have you been there?

ABOUT TWO WEEKS

Any headaches?

MY HEALTH IS FINE

Are you having memory difficulties?

JUST A FEW

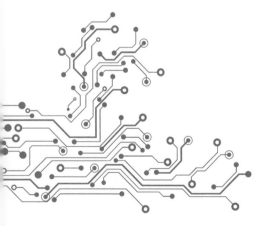

NOAM CHOMSKY
b.1928

Noam Chomsky is the founder of a modern linguistics based on formalisms, although his theories have never produced computations to model human language, a method he distrusts. He asserts the primacy of intuition as to what is a "correct" sentence and also distrusts large-scale collections of actual data, which now form the basis of language programs, such as machine translation. He claims that grammar is more important than meaning, and that all humans are born with some sketch for grammars in their brains, a claim now widely disputed, too.

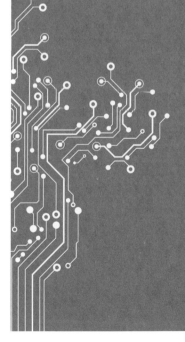

PARRY was programmed in LISP (see pages 78–81) and tested by thousands of users, who often refused to believe they had not been reacting to a human being. The program had all the interest and conversational skills that Weizenbaum's more famous but trivial ELIZA lacked. It was very robust, appeared to remember what had been said to it, and reacted badly, or even switched itself off, when internal values called "FEAR" and "ANGER" became too high. PARRY did not repeat itself and appeared anxious to contribute to the conversation when subjects about which it was paranoid were touched on, such as horses, racing, Italians, and the Mafia. It had no grammar, parsing, or logic but only a table of some 6,000 word patterns that were matched onto its input.

PARRY was quite different from AI systems driven by logic or world knowledge. It "knew" nothing—in the sense of having no stored facts—although it would have been a far better choice as a desert island companion than other AI systems of the time. John McCarthy, in whose laboratory PARRY was created, was sure that PARRY was not really AI "because it knew nothing, not even who the president is." The day after he said this, PARRY did seem to "know" that, which shows the weakness of any such criterion for "being real AI." PARRY was very close in style to the movement—started outside AI by amateur programmers—that became known as "chatbots." These were simple systems that carried on a conversation by means of tricks and ways of not quite understanding what was said, much as many people seem to do. This technology later became mainstream via the annual Loebner competition referred to earlier (see page 16) and is the inspiration behind such modern products as Siri and Alexa.

The kind of surface skimming of input that PARRY did to grasp what was being said to it, as opposed to a deep, slow grammar analysis, was very close to a later robust language technology called information extraction (IE), which now has more than fifty years of history. It could be said to have begun even earlier—at Lancaster University in the 1960s—with a program called CLAWS4, designed to do automatic part-of-speech tagging. It was the first program systematically to add to a text "what it meant"—in this case the marking of nouns and other parts of speech in the way described in Chapter 3 (see page 53). IE can locate names in text, along with their semantic types, and relates them together by structures called templates into what we could call "forms of fact," which have the structure subject–verb–object (for example, "John took a job"). These objects are virtually identical to the so-called RDF (Resource Description Format) triple stores, which are the basis of the SW (Semantic Web) in the figure on page 52; those RDF triples, too, are not quite logic, but very like IE output. As I hinted earlier, structures annotated with content like this have tended, in the AI world, to blur the distinction between human language and the content of programming languages. Also, you should not infer that researchers view that as programming languages becoming more like human languages. Some computer scientists see the reverse: that human language may be just a poor, inexact form of computer language. Many years ago, American computer scientist Carl Hewitt expressed the belief that "[human] language is essentially a side effect" of programming and knowledge manipulation.

Extensions of this technology have led to the development of effective systems to answer questions automatically from large, stored text repositories. The best example of this is IBM's competition-winning WATSON system (see pages 44–45).

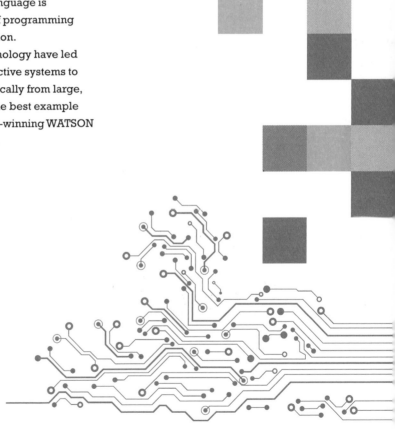

THE INVASION OF SPEECH TECHNOLOGY

The most exciting moment in recent NLP history was when it was invaded by an empirical and statistical methodology driven by successes in speech processing—taking voice input and turning it reliably into written output. The shock troops of that invasion were an IBM speech–research team under Czech-American computer researcher Frederick Jelinek, which developed a wholly novel statistical approach to machine translation. It was not ultimately successful but did better than anyone in existing MT initially expected and set in train a revolution in NLP.

Jelinek and his colleagues designed a translation system called CANDIDE that learned how to translate from English to French, and vice versa, simply from processing hundreds of millions of words of French–English parallel text, which is the form in which *Hansard*—the record of proceedings of the Canadian Parliament—is published. The choice of the project name CANDIDE, from the 1759 novel of the same name by Voltaire, was not accidental: Candide was famously young and fresh, and uncontaminated by the prejudices of the learned world. Similarly, CANDIDE did not start out prejudiced by grammar or dictionaries, or even any knowledge of the French language!

Jelinek famously boasted that CANDIDE got better whenever he fired a linguist from his team—a joke that was making a cultural point: it was a project of speech engineers, designed to extend the successes of speech recognition methods into the logic- and linguistics-dominated area of translation. CANDIDE's method was to take hundreds of millions of words of data: the long strings of French and English words that said the same thing in *Hansard*. The researchers then defined a statistical computation over the translation data that would capture the regularities in it and allow the processing of more, previously unseen, text.

The details of how Jelinek's system worked need not detain us here. The important points are that it was a series of comparisons of vast

With $p(Spanish \mid English) \times p(English)$:

Que hambre tengo yo

What hunger have	$p(s \mid e)p(e) = 0.000014 \times 0.000001$
Hungry I am so	$p(s \mid e)p(e) = 0.00001 \times 0.0000014$
I am so hungry	$p(s \mid e)p(e) = 0.000015 \times 0.0001$
Have I that hunger	$p(s \mid e)p(e) = 0.000020 \times 0.00000098$

...

numbers of words in French and English, and that a statistical value was found that selected the best equivalent in the other language for any new sentence it had not seen before. This idea of a machine being "trained" to translate, by seeing many examples of translations, is an example of what I shall explain more generally in the next chapter as machine learning.

Many were astonished that CANDIDE did as well as it did, which was to get about half the new sentences given to it more or less correctly translated. Given that the program knew no French grammar or vocabulary when it started but learned everything just from millions of words of actual translations,

↑
A statistical approach to NLP calculates the best translation across a number of possibilities.

it was a remarkable result. That it was possible undoubtedly said something about the meaning content transferred in translation and its relationship to statistical methods.

The piece of machine translation history recounted above is highly significant for our present concerns because it featured a major speech engineer (Jelinek) trying to import successful methods of speech analysis—which we will look at in a little more detail on pages 102–103—into an area that had always been considered the province of language scientists with quite different kinds of theories. It is hard to convey how shocked and horrified those scientists were that translation, a matter always considered to require fine judgment and experience, could in fact be done, for at least half the sentences in a text, not just by a machine but by one that knew no grammar, had no dictionary, and understood nothing about the other language at all. Some saw in this almost the collapse of civilization itself!

Why I described the machine translation story here is that, since CANDIDE, speech engineers have moved into human dialogue processing in the same way, claiming that their methods for speech recognition will suffice for the whole of human dialogue production. This produced something of the same shock again because it was widely felt that decoding human speech into writing does not require a great deal of information, and certainly not information about meaning.

We may all feel that we could recite foreign poetry without understanding it and, if necessary, could write it down without understanding it.

So, the idea that speech-decoding methods could perform not only machine translation but the whole of human dialogue understanding as well evokes the same defensiveness in linguists and NLP researchers not convinced about the value of statistical methods. At this point we should remember what happened in the final days of project CANDIDE: as Jelinek could not get the figures for MT above 50 percent of sentences correct and could never beat SYSTRAN in the competitions that the US government organized at the time, Jelinek reconsidered his position, as any engineer would. He began to develop techniques for building dictionaries and grammar rules of the kinds used by linguists, although he did so not by intuition, which is to say by a linguist writing those structures down for a computer to use. Instead, he developed methods so that a computer could learn dictionaries and grammar rules directly from large bodies of text. He thus gave up his distaste for what we could call linguistic structures, insisting instead that the structures be based on data and learned from actual texts. In doing this, he created a revolution in the NLP world.

Before leaving the topic of understanding language, I must add a word about the omnipresent phenomena in language understanding that I have not touched on here.

FREDERICK JELINEK
1932–2010

Frederick Jelinek was a speech scientist who spent most of his career at IBM. In the 1980s he created a high level of automatic speech recognition by statistical methods. At the end of that decade he surprised the field by claiming those methods would also suffice for machine translation trained over parallel texts. He produced reasonable machine translation without any use of linguistics or explicit knowledge of the languages. The results were never as good as other programs, but he started a revolution in the application of machine learning to language processing.

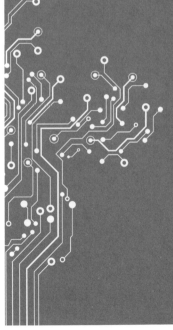

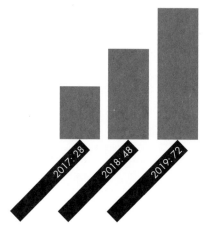

Number of countries

2017: 28
2018: 48
2019: 72

FIGURES FROM UNIVERSITY OF OXFORD

The most striking one is the *creative* use of language, of which the most obvious is metaphor, or using words figuratively. This is not a matter of poetry, as some may think, because if I take any paragraph from any morning newspaper today I shall find sentences like: "Politicians will come under increased pressure to reject the draft agreement." This is a perfectly normal, comprehensible sentence, but the sense of the verb "come under" with its object "pressure" is not one you would expect to find in a basic dictionary entry for that verb. It is figurative speech, not in this case novel or creative—rather it is close to a cliché— yet even had it been novel, as it once was,

it would have been comprehensible. Understanding such metaphors by NLP programs has always been an ongoing aspiration of AI but not one that can be considered solved. I have emphasized the practical successes of statistical and other "shallow" methods in producing working NLP systems, but this is not to imply that the decades of research on meaning representation, word senses, belief, and the intentions of speakers have been useless. I believe it is rather that we have not yet found ways of integrating them into practical systems, and this need will become apparent as the limitations of statistical and machine learning methods become clear.

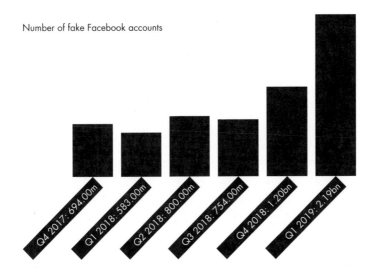

Number of fake Facebook accounts

Q4 2017: 694.00m
Q1 2018: 583.00m
Q2 2018: 800.00m
Q3 2018: 754.00m
Q4 2018: 1.20bn
Q1 2019: 2.19bn

FIGURES FROM FACEBOOK COMMUNITY STANDARDS ENFORCEMENT REPORT

NLP technology is also the basis of the production of automated Web content, a rapidly growing area whose perceived value or threat depends very much on who is behind it. Automated journalism—the writing of stories by computers from fragmentary sources, for example—legitimately profits newspaper owners. But on the darker side, the planting of computer-written reviews on hotel or restaurant websites profits those who commission them, including those who plant negative reviews on their competitors' sites. Again, NLP technologies have been good at not only writing such reviews, by sites often called "sock puppets," but also detecting them. It is, as always, a question of who has

the interest in, and resources to pay for, such detection and commissioning. Whatever its limitations, NLP is now a major industry, so let us turn to its older sibling, speech recognition, which was successful somewhat earlier than NLP.

WHAT IS AUTOMATIC SPEECH RECOGNITION?

Automatic speech recognition (ASR) by computer has made great strides and is not, at its core, a research problem anymore but simply a technology to be distributed and used. Some who work in the area might dispute this, but it is the conclusion that funding agencies, which pay for research, have drawn. The reason for this is that the methods currently being used to turn the speech waveform coming out of our mouths into comprehensible written text—a form that other computer programs could then process—may have reached a limit in their improvement, and the defects they still have may need to be remedied in other ways. Most high-end laptop computers now come with a facility for you to dictate speech and see it turn into written words on the screen—even though the words are not *understood* by the computer at all. It is for this reason that I have separated here the discussion of ASR from the earlier one of NLP, which is understanding the language the computer hears.

ASR is an automatic typewriter taking dictation, which is how IBM sells such products—as something that types what you say reasonably well after it has locked into your voice. When an ASR system makes mistakes— say, 5 percent of the time, or getting one in every twenty words wrong (and that would still be a very good system)—those errors will have to be corrected by a human, either by hand or using a spelling corrector, to make an automatic typewriter's output usable.

Experiments show that even when a recorded voice is very poor indeed—far too degraded for input to ASR—people can still figure out most of what is being said. This ability we all have is sometimes called top-down knowledge: we know what ought to be said in certain situations and we impose it, from the top, as it were, on what we hear. So, it may well be that the techniques currently used for ASR can only be improved further when linked to an "understanding" NLP system for language that knows what is going to be said.

An ASR system is anything that takes some form of electronic speech, usually from a microphone, and produces at the other end a string of words. Some systems come out with a single string of words, or what is called its "best guess." Others will provide alternatives for each word, so that some other system or a person can make choices between the alternatives because it knows what strings make sense. Remember that ASR systems do not know anything about what makes sense—they just turn sound waves into written words. So, sets of output alternatives could be like the three columns on the opposite page, where each column is a set of real words that the ASR cannot distinguish in what it has just heard.

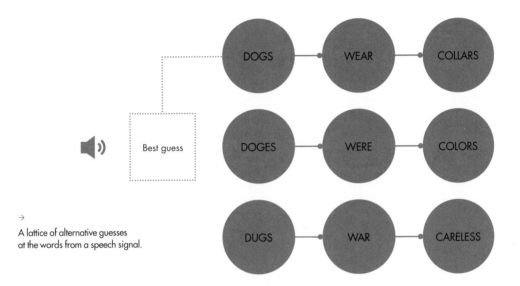

DOGS — WEAR — COLLARS

DOGES — WERE — COLORS

DUGS — WAR — CARELESS

Best guess

→
A lattice of alternative guesses
at the words from a speech signal.

In practice, these choices would be placed on a structure called a lattice, and they might have probability numbers attached, which would indicate how sure the ASR system was about each possibility. Any human could guess that the three words at the top were the most likely because "dogs wear collars" seems a more likely thing to say than any other combination we could choose. Doges in Venice also wore collars, of course, and some of them were probably careless, but we rarely talk about them, and the dogs seem much more likely overall.

ASRs are even smarter than this and can make guesses about word sequences, too, with the aid of what is called a "language model," which is evidence, gained from computing over large bodies of text, of what strings of words are more likely to occur than others. Doing that is part of machine learning, and I shall discuss it in more detail in the next chapter. It has become the central idea in modern AI, and its success in speech recognition, and the transfer of that technique to MT by Jelinek, were part of the revival of machine learning in the 1990s.

AUTOMATICALLY SYNTHESIZING SPEECH

The inverse of ASR is speech production by computer. This has always been treated as the poor relation—and as requiring less effort and research than ASR. The antique style of text-to-speech production system that English theoretical physicist Stephen Hawking continued to use from the 1980s to produce his own voice was twenty years out of date and gave the technology a bad name every time he appeared on television. In fact, there have been huge advances in the naturalness of computer speech produced from text, not only at the individual word level but at the much more difficult sentence/utterance level, where the stress in a sentence must be right or the whole thing sounds strange and false. At the word level, speech production is certainly easier than ASR in that there is no problem pronouncing by machine an unusual word such as "womb"—which is not pronounced like other "wom" words in English, as it has a long "o"—because that spelling could be inserted into the system as a special case. But the advances in speech generation in recent years have not come from special codings of difficult words but from the same kind of statistical models we have seen in ASR. Commercial systems now offer customized production systems for speaking text by computer in a range of languages, sexes, and accents.

ETHICAL ISSUES IN SPEECH

I shall discuss ethical issues briefly in Chapter 10, but one arises in relation to the speech used in domestic listening and answering devices such as Siri and Alexa, which now appear in homes everywhere. They all talk with perfectly natural voices and were initially always female. This choice has been questioned—first, because there are grounds for saying that artificial things should sound artificial, otherwise we take them to be more human than they really are and read much more into their reality than they can offer. From that view, the naturalness is a fraud.

A TEXT-TO-SPEECH PROCESS

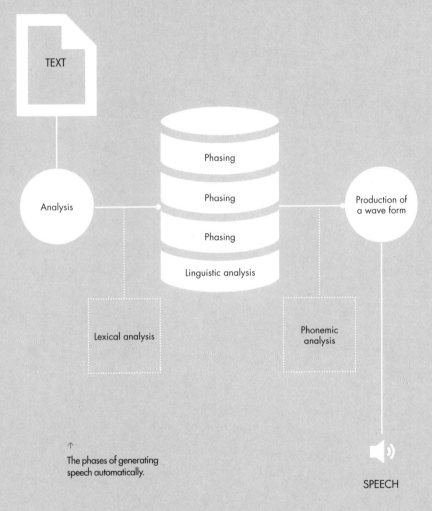

↑
The phases of generating speech automatically.

HISTORY OF VOICE TECH

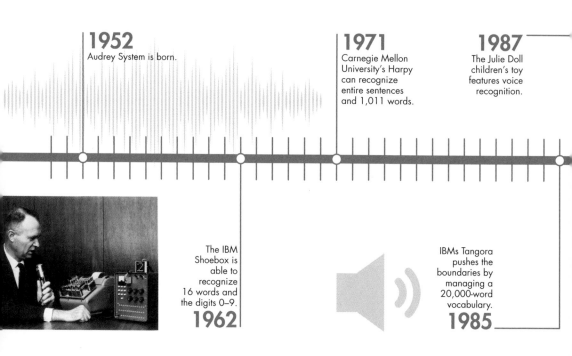

1952
Audrey System is born.

1971
Carnegie Mellon University's Harpy can recognize entire sentences and 1,011 words.

1987
The Julie Doll children's toy features voice recognition.

The IBM Shoebox is able to recognize 16 words and the digits 0–9.
1962

IBMs Tangora pushes the boundaries by managing a 20,000-word vocabulary.
1985

Secondly, you could argue that the use of female voices perpetuates stereotypes of women as servants to those who make demands of them. The latter objection is easily dealt with if the manufacturers put out both kinds of voice, as they do on GPS in cars.

The naturalness objection is more serious and brings up many questions as to how we should treat such devices, and whether how we treat them depends on how like us we

↑
A timeline of key milestones in the development of voice technology.

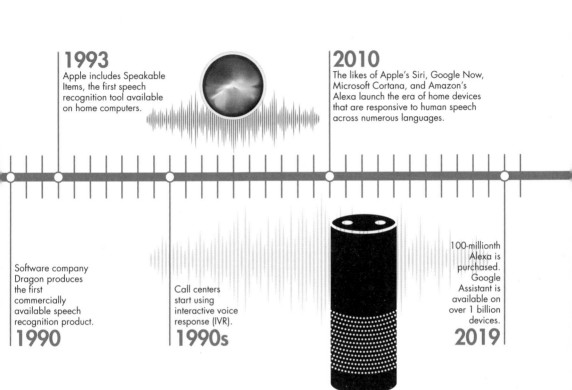

1993
Apple includes Speakable Items, the first speech recognition tool available on home computers.

2010
The likes of Apple's Siri, Google Now, Microsoft Cortana, and Amazon's Alexa launch the era of home devices that are responsive to human speech across numerous languages.

Software company Dragon produces the first commercially available speech recognition product.
1990

Call centers start using interactive voice response (IVR).
1990s

100-millionth Alexa is purchased. Google Assistant is available on over 1 billion devices.
2019

believe (even unconsciously) them to be? There has always been evidence that people tend to treat artificial things badly. If that is true, the more natural those artificial things are the better we may behave to them—and it is good to treat things well, and to teach children to do so, be they pets or robots. There is also much evidence that our treatment of others is dependent more on voice than looks. A person of a visible racial minority may well be treated better if they speak in a perfect local dialect or accent—so you might expect the same to be true for artificial speech systems. Those, such as the British speech technologist Roger K. Moore, who argue strongly for keeping "alien" Hawking-like voices in such systems, never tell us clearly what harm comes from attributing more humanity to them than they in fact have.

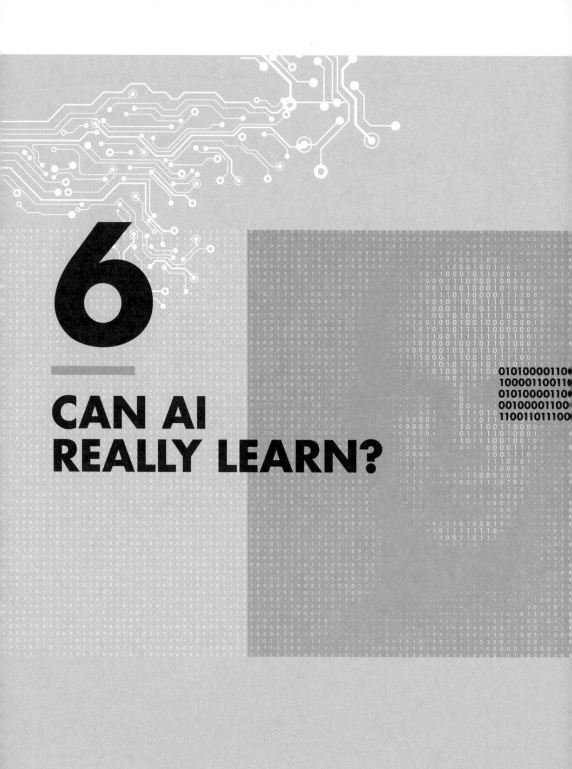

6

CAN AI
REALLY LEARN?

01010000110
10000110011
01010000110
00100001100
110011011100

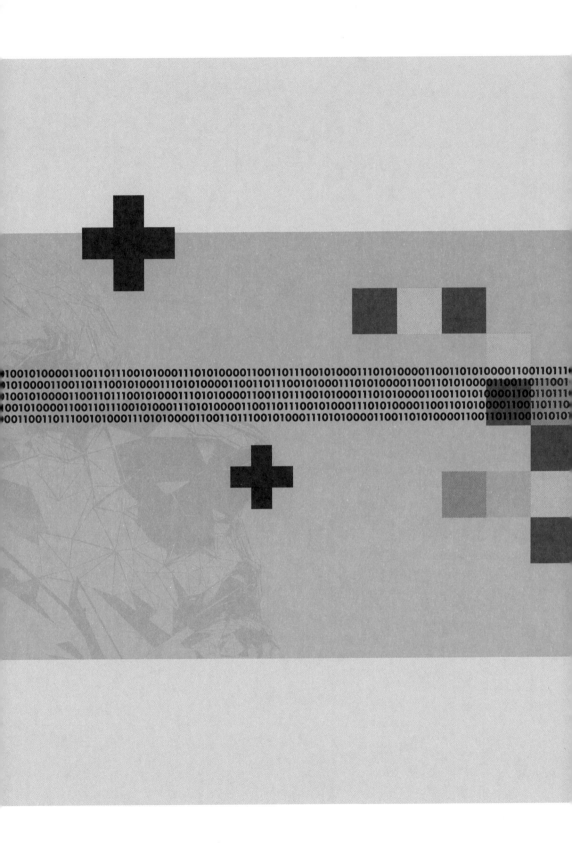

At the beginning of this book, I wrote that a modern introduction to AI could consist entirely of newspaper snippets from the week of publication, especially endless "news" about the progress of machine learning (ML). In 2019, a BBC official gave a lecture complaining that Netflix chooses what to show using "an algorithm," a choice which, she said, should on the contrary be a creative act by someone like herself. She also mentioned an ML system being used to review job applicants' CVs that was systematically rejecting female applicants.

That last comment highlights a vital point: machine learning is always based on data provided to a "learning algorithm" and depends very much on the quality of that data and how representative it is of the world.

If you give a learning algorithm lots of conversation data and ask it to learn how to reply to things said to it, then it will start to use racist or other offensive language, *if* that is the kind of material that has been used in its training. And no one should be surprised any more than they would be by children cursing after hearing that same cursing at home.

In the case of the job applicants mentioned above, if most of the successful CVs an algorithm sees come from males, then it will indeed learn that being male is a feature to look for when picking plausible job candidates.

None of the potential data confusions should, however, obscure the many real successes of ML, particularly in medicine—where, for example, an ML program managed scores as good as those of specialists in recognizing an obscure lung disease. Since your lung X-rays may be seen by people without the right skills who are also overtired and overstressed, this represents real advance. The main issue lurking behind all these piecemeal advances is whether ML, of the kinds we have now, advances any theory of *general intelligence*, over and above the striking applications I will describe.

THROWN BY THE WOLVES

Inserting even tiny disruptions into learning algorithm training materials can completely disrupt some ML processes. One particular ML program proved surprisingly good at learning to pick out wolves from dogs—which is hard for people—after being given labeled photographs of both. Later, presented with new photos, it failed. Someone then noticed that in the training photographs the wolves were always in snow. So, the program had picked out "snow" as a distinctive feature for being a wolf, until it encountered pictures of wolves against other backgrounds!

At a 2018 lecture, Australian AI pioneer Rodney Brooks asked his audience of computer scientists if they knew what "steampunk" was, and, unsurprisingly, they did not. He then showed them three pictures of people wearing retro steampunk garb, followed by a large set of images of mixed steampunk and standardly dressed people.

The audience then had no trouble picking out those with the style they had just learned from a few examples. It is precisely that human ability to learn from a handful of examples that ML cannot yet even begin to capture.

SUPERVISED LEARNING

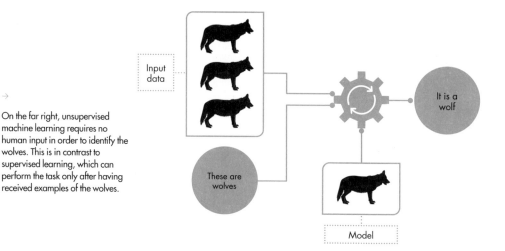

Input data

It is a wolf

These are wolves

Model

→

On the far right, unsupervised machine learning requires no human input in order to identify the wolves. This is in contrast to supervised learning, which can perform the task only after having received examples of the wolves.

I have come this far without defining "learning" because definitions often do not help understanding, but here is a distinction that will help pin down what machine learning is. The distinction is between *supervised* and *unsupervised* learning, where the first means that you give the system a lot of examples labeled with the right answers, such as pictures of wolves and dogs with those names on the images, or people in steampunk garb, so labeled. In unsupervised learning, which many feel is more like human learning, the system is not provided with such a training set of right answers. In a well-known experiment, an ML system was simply told to find out what a tiger looked like. It went to the Web and found pictures that were returned after searching on Google for TIGER, many of which would probably not have been tiger pictures at all. However, the system turned out to be good at finding what was in common

among the results and could later identify tigers fairly reliably when shown test pictures. This was an important result, because a prejudice many have against ML is that it is what is called, dismissively, "hill climbing," which is to say ML processes can only improve on, or hone, the recognition of things but can never generate new concepts they did not already know. The tiger example shows that it is not that simple. But even that may not be truly unsupervised, since the word "tiger"— though crucially not the concept—was given at the start. In an early experiment at the University of Cambridge in the 1960s, English computer scientist Roger Needham and his colleagues took a set of Greek pots and noted which had handles, rims, and images of certain types and then let the computer classify them as best it could into groups based on those features. The clumps, as they were called, of pots it discovered turned out to

UNSUPERVISED LEARNING

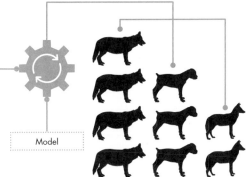

Model

be very close to the sets that the experts had already classified them into, though the experts may well not have known why they grouped them as they did, beyond intuition. That work was then called information retrieval (IR), an old technology we have already encountered (see page 27). Now we could call it early unsupervised machine learning.

The phrase "machine learning" was first used by American AI pioneer Arthur Samuel in the 1960s in relation to his program that learned the rules of the game, checkers. However, as noted in Chapter 1, early cybernetics had made learning central (though not in a digital computer) and in the 1960s, American psychologist Frank Rosenblatt designed a non-digital machine called a "perceptron," which was a simple neural net (a concept we will explore further in the next section). Another American AI

pioneer, Marvin Minsky, showed that a perceptron could not learn from examples the logical version of "OR," which is the logical function close to the ordinary meaning of "or," but which is true only when just one thing in a pair of statements is true. There is another kind of OR in logic, which is true even if both things are true. There have been decades of dispute as to whether his proof was valid, and whether Minsky knew that a more complicated neural network could have learned OR. But this "proof" of incompetence, true or false, seems to have killed off work on neural nets for some decades. It was the first AI winter, with a loss of funding and the temporary ending of AI research. The perceptron was also a transition case, like an evolutionary fossil, between the old cybernetics and the coming AI.

WHAT IS A NEURAL NET THAT LEARNS?

The world of computer networks was initially based on a metaphor drawn from the brain—consisting, like the brain, of networks of neurons. But that metaphor has not proved helpful because we know so little about how the brain works or stores information, or exactly what roles its neurons play. I can illustrate the core idea of an *artificial* neural net, and its ability to learn, by returning to a task where ML has been supremely successful: the decoding of human speech into written text.

The basic form of a neural net is a system of nodes connected by edges (represented in the diagram opposite by circles and arrows) in such a way that electronic signals of a given strength travel along edges to nodes. Then, nodes decide what to do with the signals, depending on how they are programmed. A node could multiply the strength of signal by a numerical *weight* assigned to the edge down which the signal came. The node would then sum up the signal strengths and weights it had received and decide whether to pass this new signal onward, usually depending on whether the total of weight met some threshold. As input continues, the weights can change, and this allows a node (or set of nodes) to "learn" an association between, in the current case, a sound and an English alphabet letter. Suppose a node receives, down two edges entering it, a representation of an English sound and the letter "o." Then, if the network is given many repetitions of that sound and that letter (coming in at the input nodes on the left in the diagram) it will figure out for itself ways of associating them together at the output nodes on the right, so as to give English word "transcriptions" of new sound inputs—ones it was not trained on. The sounds can be expressed as symbols called "phones," related to the symbols you see in dictionaries telling you how to pronounce new words, although in the most modern forms of learning, the networks can deal with speech waveforms (as numbers) without bothering with these phone symbols.

REPRESENTATION OF A NEURAL NET

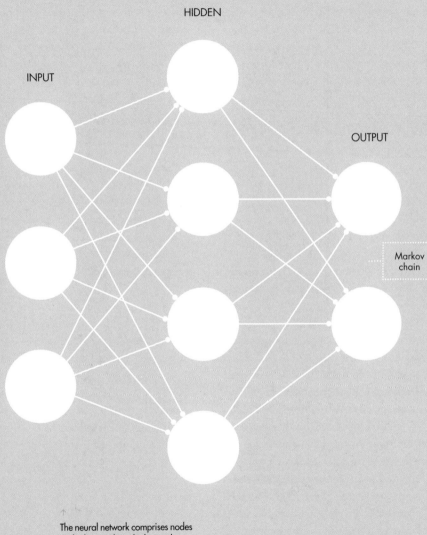

HIDDEN

INPUT

OUTPUT

Markov
chain

The neural network comprises nodes
and edges, within which signals
are processed.

This task could appear simple for English because it has an alphabet of twenty-six letters, and we are told as children that these correspond to sounds. So, surely all an automatic speech recognition (ASR) system must do is line these up and, for every incoming sound write down the letter that corresponds to it. Unfortunately, there seems to be no language that has the same number of sounds as letters, and so there is always a problem of deciding which letter to write down for a sound. And in English this problem is more difficult than in most other languages, as non-natives learning it know very well. English is one of the hardest languages to learn to spell, far harder than, say, German, whose speakers rarely make a spelling mistake.

Supervised machine learners are initially shown a very large number of sound associations, telling them that certain strings of phones are equivalent to certain strings of words. The nodes receive the sound representations for all the letters of a word, so the network must associate the sound with a letter in the context of the surrounding sounds, just as we do to decide how to pronounce one of the "wom" words in the box on the opposite page. This is called "training" and, when finished, the measure of its success is that, given a new string of phones, the network judges correctly that it is such and such a letter string and outputs it.

Two key notions at this stage are *back propagation* and *reinforcement*. The first means that when the network produces a phone-to-letter string association that is wrong—in that it does not fit the training data—that error is "propagated" back into the network. This means that the nodes that made the wrong association have the *weights* on the edges that enter them lowered, so that the wrong phone–letter association is less likely to gain as much weight in the future. All the undotted edges in the diagram on page 117 have numbers on: weights that show how likely that edge is to contribute to the output. The same basic process is followed in *reinforcement*, which comes from early behaviorist psychology. It is the notion that actions are rewarded if positive and punished if not, where that difference is expressed by raising and lowering the numerical link weights, depending on whether the phone-to-letter association produced fits the training data or not. If it does not, *back propagation* lowers the weights that caused it; if it does, *reinforcement* raises the corresponding weights on links.

Most useful systems also have one or more "hidden layers," such as the middle line of nodes in the figure on page 113; they are "hidden" only in the sense that they cannot be seen by anyone observing just the input or output layers. If there are hidden middle layers in the network, whereby the system can form clusters of associations on its own, and which we "cannot see," then the network will be able to learn better than if it went directly from input nodes (phones) to output nodes (letters). These hidden layers were the way out of the "perceptron problem" mentioned earlier that caused 1960s AI such great problems and provoked such a serious loss of faith in its capacities.

A key point for later discussion is that we could not necessarily understand the basis for such hidden associations, even if we could see them. These hidden layers constitute tools that deal only with what comes from other layers that may or may not be directly connected to the input data. Their nodes interact only with others, which may themselves be hidden. So, if the network were working to understand a visual scene—the subject of the next chapter—nodes near the input might be building up very simple notions such as the edges of an object in the scene, but hidden layers, working on those first nodes, might be building up more complex objects such as whole geometrical shapes from the edges data they had received.

CHAINS OF EVENTS

Another feature of such a neural net for ASR is that it has, in addition, a structure placed on it called a Hidden Markov Model (HMM), which is what gives it its special power. Andrey Markov was a Russian mathematician who first drew attention to the fact that sequences of things are normally not random, and you can calculate what is coming next with a certain probability based on "chains" or sequences of events seen before. The longer the chain, in general, the higher the probability of what comes next. In this case, the chains are the sequences of English letters or words as they occur in normal speech. Usually, chains of two or three letters are taken, because anything longer means too large a computation.

The extra information that a modern HMM ASR has available to it comes from what are called "language models"—information acquired from English texts on what the common sequences of English letters and words are. Huge bodies of English text are taken and a calculation is done, for every possible pair of letters in English, to establish what the probability of the next, third, letter is. This is an enormous calculation, resulting in a vast table, which says that after, say, "th" then "e" is the most probable letter (and gives a percentage probability for it) and so on for all the less probable third letters in order. The table can be consulted, letter by letter through any word, to help guess what the most likely overall sequence of letters should be, based on

the most usual orders in English. There are also language models of English word sequences, derived in the same way. And these are used to prune the output at the right side of the network based on what an output word should most probably be, given a string of words or letters that precedes it. These tables are used to adjust the output from the right-hand nodes in the diagram opposite in the following way. Think of those nodes as producing English letters in time order downward in the diagram— so that the rightmost nodes are connected by downward (dotted) links and "fire" (a metaphor from real brain neurons) one after another to produce English letters, which simply means they send something out when they receive something. The Markov sequences are used to adjust that output so it fits what is most likely in English. In the previous chapter, we saw that the following sequences could be output (as words, not just phones) from such a sequence of nodes:

Dogs	**wear**	**collars**
Doges	**were**	**colors**
Dugs	**war**	**careless**

As we saw before, the tables made up from big texts in English would certainly say that the top sequence is much more likely in English than the other two. So, if the network produces the second or third, the table will say they were very unlikely and correct them to the first. This kind of procedure does not require knowing anything about meaning, only counting how often particular sequences of words occur in texts. The procedure thus represents an obvious limitation of any statistical method in that it discounts the unlikely, which does of course happen.

This is very much how spell checks on our phones work—and we all know how annoying it is when they mistake what we want to write. Elsewhere, poetry relies on just such improbability in word sequences.

A simplified neural net in which signals enter at the top blue nodes and reach the bottom ones. Output is generated via links that have weights, and the signal strength is modified by factors that change constantly in the learning process.

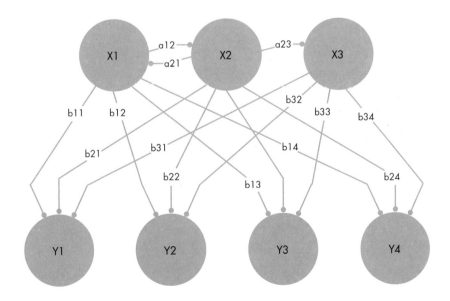

It remains remarkable that ML hugely improved speech decoding when the invented system of phonetic symbols was abandoned as a representation in the learning process. This was hugely significant for other questions, such as whether our dictionary-driven notion of word senses is useful for settling the word ambiguities that arise in machine translation, or whether human logical representations are really a good basis for machine reasoning. There is little evidence that humans use those kinds of scholarly representations in their thinking, as Peter Wason showed in the experiments described in Chapter 2 (see pages 35–37). It may be that academic studies such as linguistics lead to a wrong conception of how our languages actually function in our brains and should function in computers. A striking example of this is the Chinese typewriter that composes characters.

It was invented by Westerners, who did not have the traditional Chinese conception of how characters functioned and were composed. NLP tasks have been separated and tested and improved separately by funding agencies, eager for demonstrable progress—for example, by separating big tasks such as translation into artificial tasks such as syntactic parsing of sentences, and so on. The problem has been that those tasks have now received better statistical scores, often with ML methods, but few believe this strategy has advanced the overall task of the machine understanding of language, so necessary for real tasks such as dialogue modeling.

→

A "Double Pigeon" Chinese-language typewriter from 1974.

ANDREY MARKOV
1856–1922

Andrey Markov was a nineteenth-century Russian mathematician who studied stochastic, or statistical, symbol processes. Most famously, he gave his name to "Markov chains," which are strings of symbols that can be thought of as coming from a machine, whereby each symbol is the likeliest one to follow the chain of symbols emitted before it. Procedures like this, based on collecting data from large bodies of text, now underlie many computer applications, from generators of texts to systems that engage in dialogues with humans.

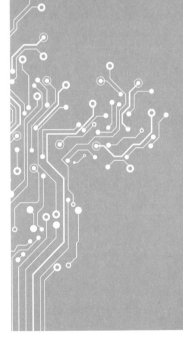

MACHINE LEARNING AND ITS EFFECT ON NLP

IN THE PAST SIXTY YEARS, THERE HAVE BEEN AT LEAST FIVE ATTEMPTS TO MAKE ML CENTRAL TO AI (EVEN THOUGH SPECULATION ABOUT THE NEED FOR IT GOES BACK AT LEAST TO BRITISH MACHINE TRANSLATION PIONEER GIL KING IN THE 1950s):

The lively state of current machine learning is due to the many reported successes of its most recent form, called deep learning (DL). This is based on the work of English computer scientist Geoffrey Hinton, who was one of the first researchers to demonstrate the use of a generalized back-propagation algorithm, with many hidden layers, for training neural networks.

Machine learning experts now see dim prospects for traditional NLP based on syntax rules and lexicons, though, as we saw in the last chapter, those have been at risk since Frederick Jelinek's work on statistical machine translation in the late 1980s (see page 94–97). "NLP is kind of like a rabbit in the headlights of the deep learning machine, waiting to be flattened," wrote a researcher recently. At the opening of his Paris ML laboratory, French computer scientist Yann LeCun said:

"The next big step for deep learning is natural language understanding, which aims to give machines the power to understand not just individual words but entire sentences and paragraphs."

This is a quote that tells you quite a bit about the understanding deep learning theorists have of NLP, indicating an apparent belief that fifty years of NLP research in AI, using many combinations of symbolic and statistical methods, have produced systems that understand only "individual words," whatever that may mean. This is a man who has never encountered Siri, WATSON, or Google Translate!

I write this not to satirize DL theorists, and I focus on them here because DL is the leading variety of machine learning, with more layers in the network than before, and it has rightly captured general interest. DL is, almost by design, detached from—

1

The failed original movement of cybernetics in the 1950s (see pages 26–27).

2

The modeling of neural nets as connectionism in various forms in the 1970s (see pages 112–113). This was an AI movement brought in by psychologists and which was the first to reintroduce neural nets. It was never able to scale up neural nets to a usable size in practice.

3

Jelinek's methods for achieving machine translation with speech–research methods in the 1980s, associating strings in two languages, as described in Chapter 5.

4

General algebraic operations on mathematical structures called *vectors*, derived from IR (see page 27). These, along with *matrices*, are general mathematical tools used to implement all modern forms of ML.

5

Nonstatistical work on machine learning as the learning of logically expressed rules of expertise. This still survives in many forms, with names such as "One-Shot Learning" and "Active Learning"—doctrines that claim to be closer to how humans learn than ML processes over huge databases that humans could never actually experience. People, this argument goes, select examples intelligently from which to learn, rather than just having masses of data thrown at them.

and ignorant of—any chosen domain field it addresses. As we saw earlier, Jelinek's team actually boasted of not knowing French when translating it—ML was to be a technology that did not require you to know anything about what you were dealing with. That might not matter if the domains tackled all shared evolutionary origins, or relationships to data, which would mean a single technique could advance them all. But this may well not be the case as written natural language has distinct properties that it does not share with, say, speech or human vision. The state we are now in with DL is a fascinating reversal of the long AI tradition that held that AI is dependent on the coding of the knowledge in the area it is exploring.

All the methods in the list on pages 120–121 had effects on NLP and AI generally and made large-scale claims that the NLP problem was essentially solved. None produced the results they promised but none of them ever entirely went away. Jelinek's methods completely redirected the course of NLP from the mid-1990s onward, even though it was the technique least likely to model human translation—because of the vast exposure to translated texts it required. All the methods hit barriers that prevented large applications in their original forms. Early connectionism could not extend beyond nets of trivial size; Jelinek's MT system never really beat the SYSTRAN hand-coded MT system in open competition; and reinforcement learning systems failed to learn anything terribly interesting.

It must be said that the limits to deep learning are not yet known. The distinguishing feature of DL, apart from having more layers of nodes, was that it did not require features on which to learn to be given to it in advance. The hope was that such systems could discover the features for themselves—as when deciding what features, like stripes, define a tiger in the earlier example—and so be truly *unsupervised*, to use the older term we defined earlier in this chapter. But it has not proved so simple to eliminate all such promptings of systems, and the fundamental discussions in DL are now about whether that elimination is really possible. The mystery of human learning means that DL programs often fail to learn from millions of examples,

although humans seem to learn new concepts from two or three—as we saw with steampunk dress (see pages 109–111)—although we have no idea how they do it.

This may suggest that written text understanding presents challenges of a kind that the techniques we have (ML and non-ML) do not yet solve, nor are likely to in the immediate future. And this is perhaps because written language has aspects of a crafted artifact, rather than a function naturally developed under evolutionary pressures such as vision and speech. Saying this is not fashionable, not least because it goes against the prejudice of most linguists that speech is primary and is real language, and its written form is secondary. The nature of the speech–language difference is harder to see from inside a language such as English, where the written form can be converted in a relatively straightforward way to a spoken form by the ASR technology we described earlier in this chapter. But from inside a language such as Chinese the issue looks completely different, since a written string can be spoken in several mutually incomprehensible ways, some of which can even be in a language from a different language family, such as Japanese! Yet understanding written text automatically is perhaps the primary goal of current AI technology.

Two other aspects of the modern ML systems mentioned above should be brought out. The key notion of *distributed representations* goes back to connectionism

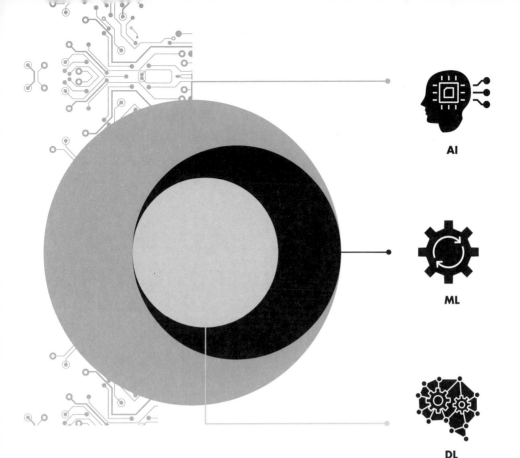

AI

ML

DL

All automated learning is within AI. Machine learning takes many forms, but within it deep learning is a modern and successful variation that can have an unlimited number of layers of nodes between the input and output nodes.

and is one of the features marking off modern ML from the older, symbolic, rule-driven forms of ML. This is the view that what we know about the brain and evolution does not support the idea that mental representations are symbolic and located in one explicit place in a brain or computer network. We should rather, the argument goes, seek representations that are distributed over structures such as networks and are implicit rather than explicit. One of the most interesting intellectual struggles in recent decades has been the effort to "have it both ways" and bring back the benefit of symbolic representations by showing that they could be represented in an implicit, distributed way. Whether this can be done is now the main issue at the heart of discussions of the ultimate capacity of DL systems.

Let us sum up the state of DL/ML today under the headings "Opacity," "Novelty," "Fusion," and "Limits."

OPACITY

A major issue now is how we are to deal with algorithms such as those of DL, whose results not even its designers and programmers always understand fully. This is not a new issue—twenty years ago, writing of an earlier generation of ML programs, American computer scientist Eugene Charniak said he would not accept the results of programs he could not understand, no matter how successful they were. Newspapers now write of "Frankenalgorithms" menacing our lives because they are incomprehensible. DARPA, the US Department of Defense funding agency, has a large research program called XAI (explainable AI) devoted precisely to getting AI to produce explanations of what ML or DL systems have produced as results, usually by running some other program that has the same effects as the DL but has a more intelligible structure. That is reminiscent of how AI was originally said to explain the brain's operations—which we do not understand well either—by simulating its results.

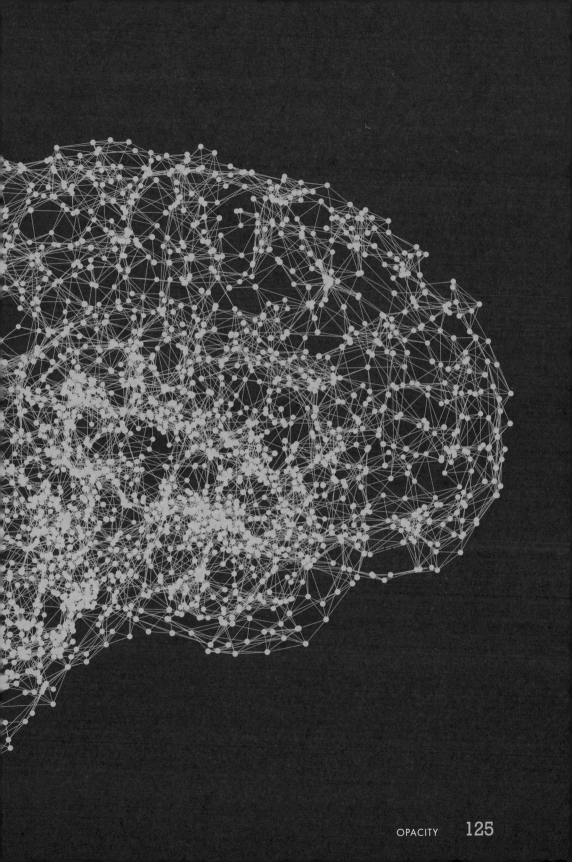

NOVELTY

1950
Alan Turing publishes his paper "Computing Machinery and Intelligence."

1958
Frank Rosenblatt develops the Perceptron, an electronic device capable of learning.

Warren McCulloch and Walter Pitts develop the Artificial Neural Network.
1943

Marvin Minsky and Seymour Papert publish *Perceptrons: An Introduction to Computational Geometry.*
1969

A familiar trope in the history of AI is the rediscovery and renaming of earlier work, or at least the exaggeration, for fundraising purposes, of the differences between earlier work and something new. This is partly due to a lack of scholarly memory in AI and the relentless emphasis on novelty and forward movement. It is also, more benignly, because the stunning advances in hardware processing have made computations possible that could be imagined in the past but not carried out on the computers of those earlier times. In the case of DL, it is reasonable to ask whether it differs all that much from earlier statistical processes with their access to large data and other forms of ML using neural nets and statistics.

A timeline charting the development of machine learning and deep learning through a number of important innovations.

DEEP LEARNING TIMELINE

1980
Professor Teuvo Kohonen develops the Self-Organizing Map.

2006
Deep Boltzmann Machines are developed.

Deep Belief Networks emerge.

2012
The concept of dropout improves deep neural networks.

The Restricted Boltzmann Machine is created.
1986

Building on the Recurrent Neural Network (RNN), bidirectional RNNs increased the input available to a network.

The Generative Adversarial Network (GAN) is developed.
2014

The Boltzmann Machine is promoted in machine learning.
1985

Long Short-term Memory (LSTM) is developed.
1997

The Capsule Network emerges.
2017

FUSION

Much of the current argument and dispute surrounding ML/DL concerns the vexed issue of how far a learning system can find "classifiers," "priors," or "features" of the world for itself in the data it is given, or how far these have to be smuggled into the design by the researchers or implementers. This is the more modern version of what I touched on earlier when describing Jelinek's revised machine translation program (see pages 96–99). There he moved from simple statistical associations of sentences in the two languages to methods for statistically creating linguistic forms of knowledge such as grammars and dictionaries, thus priming his system by building knowledge of the (linguistic) world into it to improve its performance. It is also the problem of stopping the wolf detector choosing "being in snow" as the defining feature of wolves (see page 109). There is always a risk that ML/DL systems will hit upon features with which to classify that we not only cannot understand, but which are absurd.

↑
AI can be said to divide into a number of branches, of which machine learning is one.

The positive way of looking at this situation is that the feature an individual node classifies may be vague, but part of the magic of AI has been to bring together vague constraints to make strong ones. That was always a feature of classical AI and is reappearing in the age of ML. The current internal debate is also about whether an ML system based on statistics can also benefit from the decades of work on reasoning and logical representations.

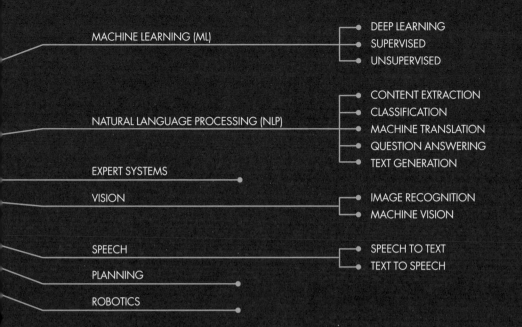

MACHINE LEARNING (ML)
- DEEP LEARNING
- SUPERVISED
- UNSUPERVISED

NATURAL LANGUAGE PROCESSING (NLP)
- CONTENT EXTRACTION
- CLASSIFICATION
- MACHINE TRANSLATION
- QUESTION ANSWERING
- TEXT GENERATION

EXPERT SYSTEMS

VISION
- IMAGE RECOGNITION
- MACHINE VISION

SPEECH
- SPEECH TO TEXT
- TEXT TO SPEECH

PLANNING

ROBOTICS

FROM THE VAGUE TO THE CERTAIN

An anecdote ascribed to Geoffrey Hinton makes the point that you can combine vague inputs to make something definite, even if one of those inputs might be false. Ask yourself who X is in the following vague sentences:

1   X was an actor.

2   X was extremely intelligent.

3   X was a US president.

Know who this is?

Hinton assumed many would guess Ronald Reagan because he was the only US president who was an actor and that was more significant than the, again assumed, falsity of (2).

LIMITS

A crucial issue is whether there are limits to the reach of DL. Israeli-American computer scientist Judea Pearl has argued strongly that the associations on which deep learning is based cannot grasp the key concept of *cause*, which is to say of one thing causing another versus its being associated or correlated with another. This is one of the oldest philosophical problems, going back to Scottish philosopher David Hume in the eighteenth century, and still preoccupies much of medical research. The British philosopher Bertrand Russell used to give the example of factory hooters going off in Manchester every day at 5pm and workers then leaving factories in London. There was complete correlation but clearly no causation because they could not be heard 200 miles away. Pearl is basically aligning himself with the philosophers in arguing that causation is never in the data of the world, rather it is something we impose on it in order to understand. DL/ML mechanisms cannot derive any such notion and can never therefore really understand the world.

But none of these doubts can take away the extraordinary successes of DL in such things as playing the game Go, and in speech and face recognition.

The claims of DL—by which I mean separately trained layers of networks, of the kind advocated by Geoffrey Hinton—have not been proved successful yet for NLP (such as conducting coherent and extensive dialogues), but they may yet do so. The fact that we have been in positions very much like this before with earlier ML methods, as I noted above, and that those earlier claims were never substantiated, does not mean they will not be this time around. There have been real advances in recent decades and not everything is a rediscovery of the obvious.

Most recently, networks with millions of nodes—exactly what eluded earlier connectionism—have been created for language computation, and have produced prototypes with names such as GPT3, which create extraordinary prose of seemingly humanlike quality and relevance—and even

one or two quite good jokes. Perhaps the most striking demonstration of the reach of the GPT3 prototype was an exercise where a version of the network with access to the writings of the Australian philosopher David Chalmers engaged in a dialogue with him, effectively producing a plausible dialogue with himself. Such networks have even begun to choose correctly between alternative explanations of everyday situations, and some believe some kind of artificial general intelligence is very close.

I personally remain skeptical, if only because the successes of NLP in recent years, such as Siri and WATSON, owe little to ML so far. But we shall see.

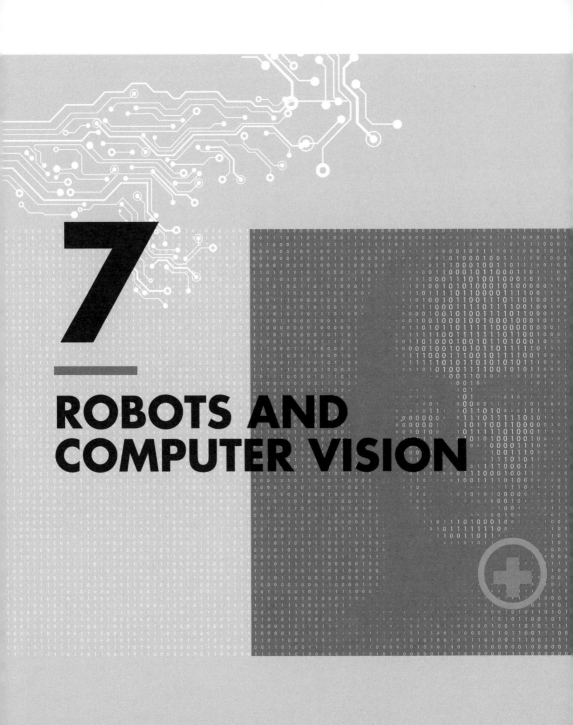

7

ROBOTS AND COMPUTER VISION

In this chapter I will look at both robots and
computer vision, where the latter is the technology
of getting computers to recognize things such as
human faces. These have usually been taken to be
the more "engineering" parts of AI, less concerned
with modeling how we humans carry out what we
do than getting tasks done. Robots are still for many
the stereotypical form of AI. When people think
of the subject, it is robots they see, as do most
cartoonists, who usually portray clunky things
in the shape of human bodies.

On the right, a "bird" drone created by Stanford University engineers, which combines four rotorblades and birdlike clawed feet.

On the left, robotic vision sensors such as this are being integrated into automated industrial processes.

Robotics and computer vision have historically sat outside the intellectual core of AI. First, this has been because they called so much on other technologies—on screens, on mechanical engineering, on pattern-matching, and on control engineering. Secondly, robotics and computer vision were more separate from mainstream AI because they had little use for what I have described as its logical or representation preoccupations and were always much more concerned with mathematics and statistics. They were, you could say—using a key distinction in mathematics—always concerned with values that vary continuously, and could take any numerical value, which logic normally cannot, because things are either true or false in most logics.

We are taking the two technologies together because they have traditionally been associated by their differences from the rest of AI and have often been combined. They do not strictly have to be present in the same concrete forms—face recognition cameras do not need arms, and though autonomous cars and drones have vision, car-making robots do not. For most of those working in robotics, however, machine vision is not a separate subject but part of robotics.

ROBOTS

The word "robot" derives from the Czech for "forced labor" and first appeared in its modern sense in Karel Čapek's play *Rossum's Universal Robots* in 1920. The idea, however, is ancient, and Roman, Greek, and Jewish civilizations all speculated about the possibility of artificial and mechanical persons—though without electricity it was almost impossible to make one. But why is it that I still think of humanlike robots first, usually from movies such as *The Terminator*? Is there any need at all for robots to look like people? Few ask what it is that we would want human-shaped robots for. An answer that comes quickly to mind, and fits the *Star Wars* movie genre, is as artificial soldiers. But that may not make much sense in the real world— why would they be better than tiny robot tanks with guns or flying drones with rockets? Matching biological equivalents does not always make sense. When the US Army commissioned robots to carry soldiers' backpacks over rough terrain, the most successful products were six-legged things

↑
The robots used in car-assembly plants are descendants of Stanford's pioneering robotic arms. They are able to move in three dimensions and can regulate their actions to ensure safe and effective operation.

foot in the way we do. There are humanoid
robots that do extraordinary somersaults from
a standing position. These are engineering
marvels, even though most robots now in
productive use are articulated arms that
assemble cars in factories, and which have
little similarity to the humans who used to
do that job.

like large insects, rather than anything
resembling the real-life mules or horses that
used to do the job for the military.

Nonetheless, and whether the human
shape has its uses in robots or not, some
humanoids have done remarkable things.
Robots from the American company Boston
Dynamics climb stairs easily and cannot be
pushed over, moving instead onto the back

There can be no fixed constraints on
the form a robot can take as long as it is
computer-controlled and part of it moves
freely, even if the remainder does not (as in
car-assembly factories). An automated car
is increasingly a robot itself—a powerful
computer that happens to have found wheels.
So is a drone, with neither wheels nor arms.

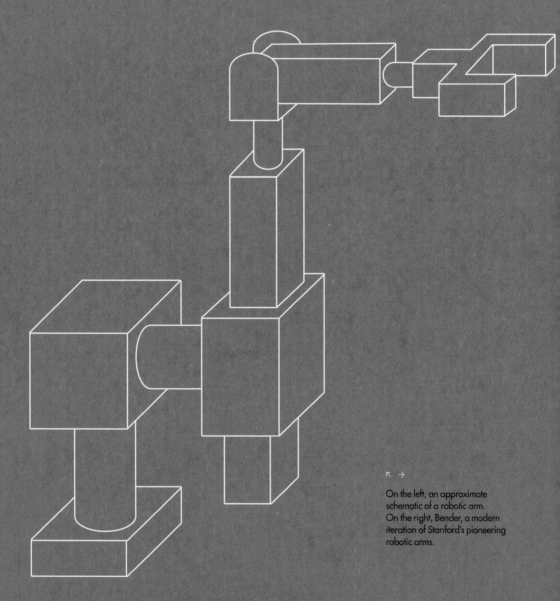

↖ →

On the left, an approximate
schematic of a robotic arm.
On the right, Bender, a modern
iteration of Stanford's pioneering
robotic arms.

A HELPING HAND

Assembly-line robots are all descendants of what, at Stanford in the 1970s, was called the Scheinman Arm, invented by the American robotics pioneer Victor Scheinman. This was the first computer-controlled arm that could move freely in three dimensions, that had fingers, and used "feedback"—a notion we met in the previous chapter as back-propagation in a network—from the fingers so as not to close too tightly when holding an object. Scheinman sold his design to Unimation, who then teamed with General Motors to develop the robotic manufacture of cars—and the rest is history.

Robots in the future will go to places we cannot, such as down miles of narrow sewage pipes to check them, or through our bodies—where they could not possibly be shaped like us. And yet, there will be uses for humanlike robots, such as the following obvious examples.

> Exoskeletons to make our bodies more powerful, including "magic trousers" for those who cannot walk after spinal injury.

> Household servants—and perhaps most robots that will work in human environments—that must be near us much of the time, and even climb stairs.

> Sex robots—inevitably—since the sex industry is always a larger part of any technical development than is generally recognized.

The third technology in the list is still in its infancy—and most sex robots are still a little pathetic—but there is no doubt the market will grow as the technology improves. This is already creating a serious ethical debate about the treatment of artificial persons, and I shall turn to the discussion of Artificial Companions in Chapter 8. British chess master and futurologist David Levy has argued strongly that this future market will not just be about sex, but about affectionate relationships with artificial people, replacing human ones. Some future robots will also undoubtedly be warriors and their autonomy has also created a lively ethical debate that I will cover in Chapter 10.

Human-shaped robots have been a staple of TV and movie fiction, and realistic ones can of course be played by real actors, which has created a new style of screen performance. But as yet little is known about how people will react when, and if, such entities become commonplace. There are a few isolated and interesting results already, such as the fact that people seem to prefer such robots to approach them from the side rather than head-on. At present, though, autonomous cars are the likeliest form in which robots will present themselves to us in everyday life, and it is significant that their introduction is happening gradually, via the assumption of more and more complex tasks.

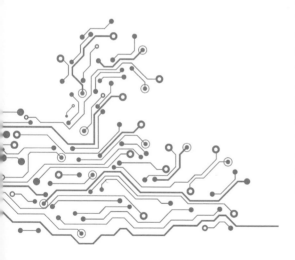

Production model cars have had cruise control for decades, but now some offer automatic reversing into parking spaces, retrieval from underground car parks, and automated lane-following on superhighways. Soon, we will even have convoys of trucks moving under computer control. This gradual introduction of functionalities is a good way of getting the public used to relinquishing control to automation in stages; and it reflects the fact that most automation of driving is not AI in any traditional sense, but a set of localized engineering solutions to particular problems, such as negotiating traffic circles. An automated car does not have to plan on any larger scale—the journey plan is already subcontracted to the GPS system and not generally part of the car. When you take a wrong turning, the GPS must calculate how to get back onto the original route as quickly as possible, which is a really striking piece of AI of the classical type. In the next chapter, I shall discuss Artificial Companions, which can help plan our lives. Eventually, there will be no barrier between these and the automated car robot—the car will itself be one manifestation of our "companion," and then the car will indeed have these planning and knowledge capabilities.

Most robots, whether car builders or in human form, need to have some classical AI content, to enable them to plan tasks and know when they have succeeded. In the 1970s, the Shakey robot developed at Stanford Research Institute pioneered a way of conceiving plans in a form called STRIPS (Stanford Research Institute Problem Solver). This was really a collection of simple scripts that specified the order in which tasks should be done, rather than working that out in a more complex way as in standard planning theory. Robots are a continuation of development from the older cybernetics into mainstream AI, and both are embedded in engineering and continuous processes. Another strand in robotics is that represented by the work of Australian roboticist Rodney Brooks, called behavior-based robotics. This focuses on modeling simple organisms, including ants, and their organizing capacity without any apparent notion of humanlike reasoning. Very much a continuing influence of cybernetics, it has resulted in moderately successful products such as Roomba®, an automatic vacuum cleaner that moves freely round a room.

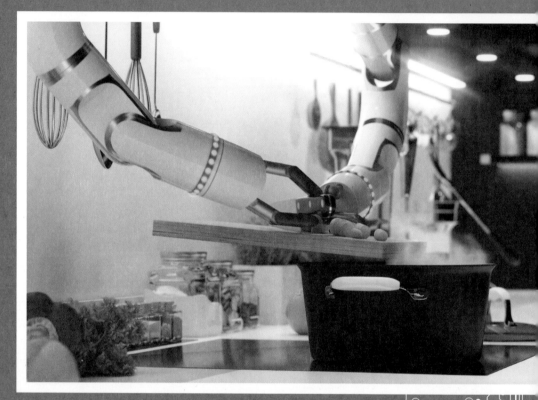

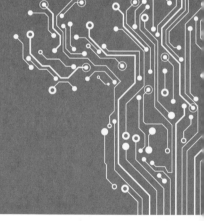

↑
A three-dimensional illustration of
a robotic kitchen assistant—just one
possible application of robotics in
the domestic environment.

↑
Exoskeletons are wearable
machines that are designed
to enhance human mobility,
strength, and endurance.

COMPUTER VISION

I have referred several times to the classical AI paradigm as logic-based, but it was never entirely devoid of numbers because AI theories of vision flourished close to pattern-recognition research, which was entirely statistical. Computer vision, always an engineering discipline, had numbers from the beginning. The magnitude of the computer vision task was enormously underestimated in early AI—there were suggestions in the 1950s that it was a graduate student project with a camera and a computer! Even now, there is not complete agreement on the task's basis, despite the tremendous advances in facial recognition, along with vehicle/pedestrian/bicycle recognition in automated cars.

The opposition I have described within every part of AI between symbolic representations and statistical/neural ones (where representations are absent or secondary) also crops up in computer vision. In its earliest days, there were claims by American computer scientist David Waltz and others that there was a "grammar of vision" that would enable scenes to be represented in a computer with rules like those of a linguistic syntax, although this approach was not successful.

At every stage, computer vision has advanced by dealing with a concrete problem or application rather than concentrating on a general theory of "how we see." A small illustrative example is the recognition of handwritten numbers and letters. Optical character recognition (OCR) of printed fonts has been very successful, but freehand-written characters have been resistant to computer recognition. When zip codes were first introduced to the UK in 1959, it was assumed that their letters and numbers would soon be recognized by machines, so that sorting of letters could be done automatically. This has never happened and the codes are still typed on by hand to enable later machine sorting by OCR. Yet in the old USSR, where there was only one type of envelope, this problem was solved by a set of boxes in the bottom corner of every envelope with dots

Computer vision comprises an arrangement of pixels—seen here as blocks of various shades.

Each pixel's color is assigned a number. These numbers typically range from 0 to 255 and can be thought of as forming a grid.

159	156	160	161	166	134	128	136	141	159	162	149
159	157	163	167	106	77	94	109	97	67	148	152
159	160	167	97	37	54	96	110	85	66	130	157
156	158	165	30	30	76	116	128	112	135	123	156
158	161	157	161	192	205	214	220	219	174	97	138
160	164	150	116	182	211	218	219	217	162	54	147
164	165	148	92	93	193	105	121	179	180	96	166
163	165	169	189	199	221	195	178	189	179	173	155
158	165	189	149	185	216	222	213	196	167	168	
153	161	160	147	154	202	214	208	207	208	174	171
151	159	167	160	143	168	190	202	196	171	170	172
149	160	142	130	190	214	205	193	184	149	177	174
127	80	35	163	118	123	175	209	202	106	118	162
26	22	29	150	215	132	203	242	174	108	98	89
25	28	33	132	185	99	187	236	130	108	95	81
24	33	37	89	150	89	218	175	111	97	87	89

Here is the grid of numbers with the picture of Alan Turing removed. This is how a computer would interpret the image of Turing.

↓

The machine learning process a computer follows in order to "see" can be described in four stages.

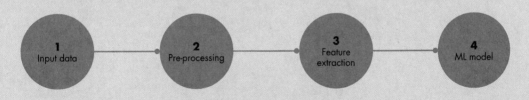

1
Input data

2
Pre-processing

3
Feature extraction

4
ML model

→

In the USSR there was only one kind of envelope and zip codes had to be written in slots with a pencil connecting dots. This made automatic recognition of zip codes easy.

that a sender could connect to make stylized zip code numbers. Machines recognized these long ago because the problem had been constrained to a simple form by central social control so that it was solvable—unlike the individualistic written characters in the UK. There was a clear moral in this story: like all AI, machine vision is always more likely to work for precisely defined and controlled tasks.

Typical tasks within computer vision include scene reconstruction, event detection, video tracking, object recognition, and image restoration (such as enhancing telescope images of planets). These are all concerned with analyzing visual input to the computer, normally via a camera but possibly a radio or other telescope. There is now also the huge field of computer graphics, including the production of movies without "natural" scenes, and increasingly the special hardware developed for that has driven the analysis tasks as well. The detection of events in video has been a more recent preoccupation; and it

is of great interest to security agencies to detect the interaction between individuals in public, or to be able to detect those people simply loitering!

The classical problem in computer vision, however, was always that of determining whether incoming image data contained a scene that the computer could reconstruct and then search, to find some specific object or feature. These can be very specific indeed, such as in the considerable research now going on to detect pornographic images in public media, usually by searching shapes in flesh tones that take up much of an image. Again, the recognition of human faces has always been a major goal, and the only one that may reflect innate properties of the brain, since there is good reason to think the human face may be the only thing a newborn recognizes without learning it. The vital feature in recognizing a specific human face now seems to be a ratio of measurements across the eyes, mouth, and nose, and this may be as distinctive of an individual as a fingerprint. Nonetheless, and despite great advances—for example, face recognition by CCTV cameras in public places—it is not yet anywhere near 100 percent accurate.

This classic phase of computer vision was, as in the case of other areas of AI we have discussed, the building up of a representation from the data out there—detecting straight edges in a scene or working with the texture of the surface of an object that could be made to indicate its geometry. The goal was to mimic what humans were believed to do with their eyes and brains—to build up a three-dimensional representation of what was in front of the camera from the two-dimensional, flat array of pixels of light, dark, and color that a camera detects. It was assumed that the visual data is in some sense impoverished—poor in information—and that the brain or computer has to put in much of the information structure that we find in a visual scene. This belief goes back to eighteenth-century philosophy, particularly the ideas of German philosopher Immanuel Kant, which worked their way to AI via psychology. These state that the structure of space (and time) is not simply *out there in the world* at all but supplied by our understanding, because without our doing that the world is simply not comprehensible.

The key figure who crystallized this view in AI was English psychologist and AI researcher David Marr, who believed that AI vision must create what he called a "two-and-a-half dimensional sketch" from

this poor camera data and then from that create a full three-dimensional representation with which to understand what it was seeing. A long-running debate has ensued between supporters of Marr (who died too early to take part) and the American psychologist James J. Gibson, who held that this representation-centered view did not explain how different species seemed to handle vision quite differently and in ways that could only be understood in terms of their lifestyles—such as catching small, fast prey or avoiding big, slow-moving enemies. Gibson argued that there was no reason to believe these species all built up representations in the way Marr had described, but rather that they detected directly what they needed in the information coming to them. This dispute is partly a matter of emphasis, because in his great book *Vision*, Marr had written: "Vision is used in such a bewildering variety of ways that the visual systems of animals must differ significantly from one another."

Yet there is real difference here in belief about the degree to which a human or machine visual system should provide, in advance, much of what we see in the world, and the degree to which that *a priori* element is in the representation or, as Gibson, suggested, implicitly built into the perceptual machinery itself. This dispute came via philosophy and turned into one between two psychologists about how humans see, and yet its implications for AI

← An early automated taxi in California.

were profound. As in the other areas of AI that I have discussed, such disputes about theory have been largely bypassed by advances in machine learning. It is now generally accepted that a special class of neural nets called *convoluted neural nets* provide the best results in visual recognition and do so without taking sides in the Marr–Gibson debate, though any approach without representations, as neural nets are, must be seen as leaning toward Gibson.

The example of road signs shows how separate different AI functionalities still are. Automated cars have already traveled millions of miles on public roads and recognize and obey standard road signs as they do so. Yet in Web security tests—usually called Captcha—to ensure that you are human and "not a robot," as the websites put it, you have to pick out which of a set of pictures include a road sign, on the assumption that a robot "Web crawler" could not do that.

The important emphasis in ML approaches to computer vision and their success is specificity. An automated car has a large set of vision systems all around it looking for very specific things: for pedestrians, for bicycles, for cars of a certain size, and so on. It will not recognize a taxi or a bus unless it has been trained to, and this is an utterly different way of looking at computer vision than the old philosophy- and psychology-motivated belief that we build up large representations of all we see.

I have contrasted the modeling of specific low-level skills, such as recognizing a particular kind of object like a bicycle, with a much more general embodied intelligence that we may hope to see in automated cars, one that will know our habits, needs, and usual destinations. This is the notion of the personal Artificial Companion and we shall now turn to that important notion.

DAVID MARR
1945–1980

David Marr was a twentieth-century British scientist involved in the neurophysiology of human visual recognition. His work had great influence on the development of visual recognition by computers. He proposed the audacious idea that our brains construct a three-dimensional environment from the basically flat, two-dimensional scenes that human eyes see. Marr posited an intermediate state called a two-and-a-half-dimensional "primal sketch," from which a full representation could be derived and which algorithms could construct from the lines and edges they "saw" initially. This work is now considered to have been superseded in AI by machine learning methods.

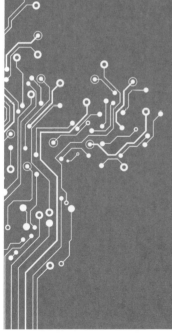

8

MAKING THINGS PERSONAL: ARTIFICIAL COMPANIONS

01010000110
10000110011
01010000110
00100001100
11001101110

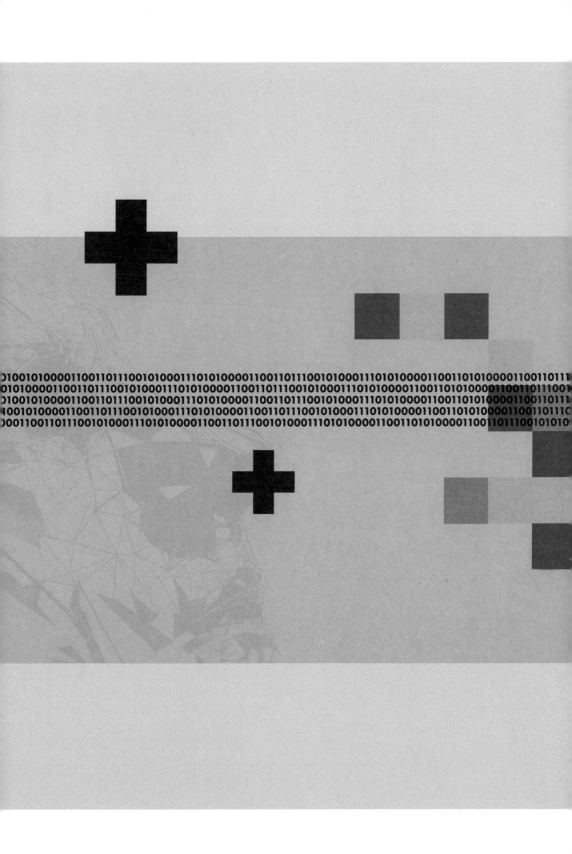

**We can be quite sure that Artificial Companions
are coming. In a small way they already arrived in
1996, and millions of people have already met them.
The Japanese Tamagotchi (literally "little eggs") were
a brief craze in the West that saw sensible people rushing
home to play with their toy so it would not pine, where
"pining" meant sad eyes and icons on a tiny screen
and playing with it meant pushing a feed button!
The extraordinary thing about the Tamagotchi (and later
Furby, in the US) phenomenon was that people felt guilt
about their behavior toward a small cheap toy that
could not even speak.**

The Tamagotchi episode said a great deal
about people and their ability to create and
transfer their affections to inanimate objects.
This phenomenon is almost certainly a sign
of what is to come and of how easily people
will find it to identify with, and care for,
automata that talk and appear to remember
who they are talking to. This chapter is about
what it will be like when much more
sophisticated objects are available—and
since most of the basic technologies, such
as speech recognition and simple machine
reasoning and memory are already in place,
this will not be long. Simple robot home
assistants are already available in Japan,
but we only have to look at their hard plastic
exteriors and listen to their tinny phrases
about starting the dishwasher to realize
that this is almost certainly not how an
ideal Companion should be.

↓

The Tamagotchi was an early
and primitive Japanese toy that
had no language but had to be
"fed" by pushing buttons.

An early Companion mobile robot: Astra from Amazon.

COMPANION SCENARIOS: SENIOR, JUNIOR, AND DRAGOMEN

Companions should not be about how to fool us, as in the Loebner competition I discussed earlier (see page 16), because they will not pretend to be human at all. Imagine the following scenario: an elderly person sits on a couch, and beside them is a soft toy or a large furry purse, which I shall call a Senior Companion. It is easy to carry about, but much of the day it just sits there and chats; it will explain the plots of the TV serial they are watching if forgotten and will know all about the children and grandchildren. Given the experience of Tamagotchi, and the fact that elderly people with pets survive better than those without, we can expect this to be an essential lifespan- and health-improving object to own. The elderly population of the EU and the US is the most rapidly growing segment of the population, and one relatively well provided with funds to buy technology.

A large proportion of today's elderly are effectively excluded from information technology, the Web, the Internet, and some cell phones because "they cannot learn to cope with the buttons." This can be a generational issue or because of loss of skill with age. There are talking books in abundance but many otherwise intelligent elderly people still cannot manipulate a TV remote control, which has too many small buttons with unwanted functionalities. All this is well known, and yet there is little thought given to how our growing body of elderly people can have access to at least some of the benefits of information technology without the ability to operate a device.

After all, their needs are real and pressing, not just to have someone to talk to, as they increasingly live alone, but to deal with communications from public bodies, such as councils and utility companies demanding payment, and to set up times to be visited by nurses or relatives. Other needs are how to be sure they have taken their pills, when keeping any kind of diary may have become difficult, and ordering foods when the delivery services available via the Web are difficult for them to use.

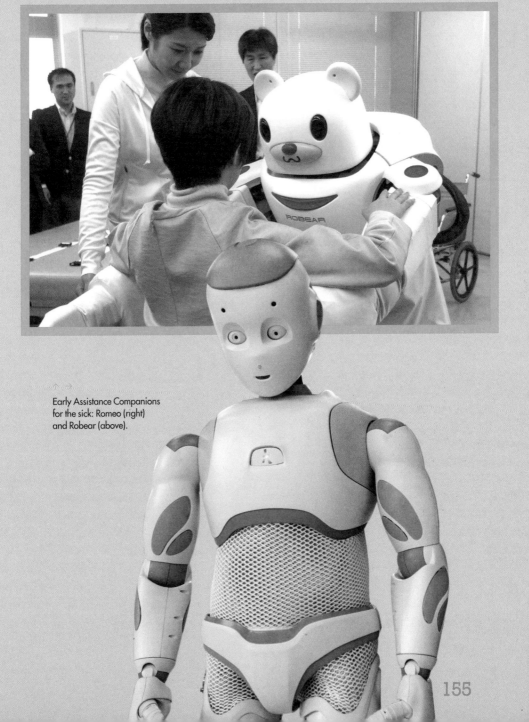

Early Assistance Companions
for the sick: Romeo (right)
and Robear (above).

155

A Senior Companion that could talk and understand, and gain access to the Web, email, and a cell phone, could become an essential mental prosthesis for an elderly person, one that any responsible society would have to support. But there are also aspects of this beyond just getting information, such as having newspapers enlarged on the TV screen until the print is big enough to read. It is these needs that products such as Siri and Alexa have begun to meet in a limited way.

Many elderly people spend much of their day sorting and looking over photographs of themselves and their families, along with places they have lived and visited. This will obviously increase as time goes on and everyone begins to have access to digitized photos and videos of their whole lives. We can see organizing this as an attempt to establish the narrative of our lives, and to make sense of them. It is what drives the most literate segment of the population to write memoirs (if only for their children) even when, objectively speaking, they may have lived lives with little to report. But think of what will be needed if a huge volume of digital material is to be sorted, of the kind that everyone now acquires.

We can see all this as democratizing the art of memoir. It will mean far more than simply providing ways in which people can massage photos and videos into some kind of order on a big glossy screen—it will require a guiding intelligence to provide and amplify a narrative that imposes a time order. Lives have a natural time order, but this is sometimes difficult to impose and recover for an individual. Even those with no noticeable aging issues find it hard to be sure in what order two major life events actually happened, for example:

"I know I married Lily before Susan but in which marriage did my father die?"

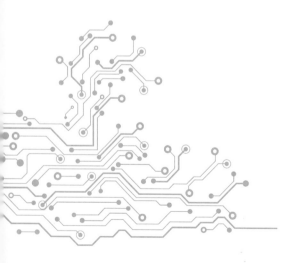

Such an example illustrates how an artificial agent might assist in bringing the events of a whole human life, whether in text or pictures, into some coherent order. This is the kind of thing today's computers can be surprisingly good at, but the time ordering of events is a complex and abstract notion, which can be simple ("I know James was born before Ronnie") or comprises only what is called a partial order in other situations ("I know my brother's children were born after my marriage and before my wife died, but I'm not sure in what order they came"). Issues like these represent real problems at the border of memory and reasoning for many people, especially the elderly. As well as seeing the Companion as a platform embodying many of the functions present in the central AI project, it also represents a real task, one where success, even partial success, would be a boon to millions.

Other Companions are just as plausible as the Senior version. A Junior Companion for children would most likely take the form of a small and hard-to-remove backpack that always knew where the child was, saw them safely across roads, and conversed with them in elementary French, perhaps, on the way to school. Or consider an artificial dragoman— the old Ottoman interpreter and guide for travelers—who, on vacation, would not only translate and guide you to the best sites, but could sit and converse on your behalf with foreigners you encountered, showing them just the right pictures of your children and displaying theirs to you.

COMPANIONS AS A CORE AI PROJECT

The much advertised Siri and Alexa are the first commercial attempts to get talking Artificial Companions into our lives, but the notion has been around for decades, as with everything in AI. Companions are a crucial incarnation of AI because they embody so much of what is central to the AI project:

language, understanding, reasoning, empathy, planning, and so on.

Some Companions might be robots in the future, but there is no need for human form, as opposed to a toy or a phone. I will be concerned here with aspects of Companions such that embodiment is a secondary matter, provided they can converse with an owner and can reach out to the world via the Internet for information and to establish action and control.

I will distinguish Companions from conversational Internet agents that carry out specific tasks, such as railroad and plane ticket ordering applications with speech dialogue. Those go back to early MIT systems in the 1990s and have now evolved into such things as intelligent talk-to microwaves and TV controllers without buttons. But I shall mean something more general by Companion—an entity that in principle knows all about you.

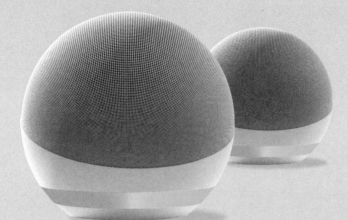

← ↑

Apple's Siri (2011) and Amazon's Alexa (2014) were pioneers of the emerging market for domestic talking Artificial Companions.

FIVE DISTINGUISHING FEATURES OF A COMPANION AGENT

I take the distinguishing features of a Companion agent to be that:

1 It has no central or overriding task, and there is no point at which its conversation is complete or has to stop, although it may have some tasks it carries out and completes in the course of a conversation.

2 It should be capable of sustained discourse over a long period—possibly the whole lifetime of its principal user.

3 It is essentially the Companion of a particular individual, its principal user, about whom it has a great deal of personal knowledge, and whose interests it serves. It could, in principle, contain all the information associated with a single life.

4 It establishes some form of relationship with that user, if that is appropriate, which would have aspects we associate with the term "emotion."

5 It is not essentially an Internet agent or interface, but since it will need to access the Internet for information (including everything about its user and their media use and searches) we could take it to be a kind of Internet agent.

KEY QUESTIONS ABOUT
COMPANION AGENTS

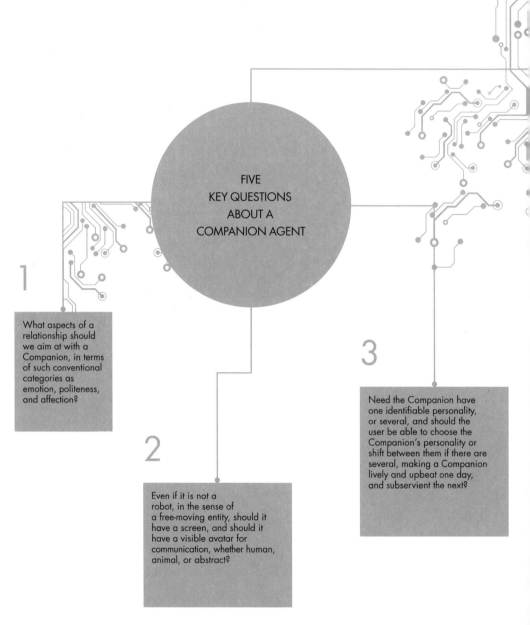

FIVE
KEY QUESTIONS
ABOUT A
COMPANION AGENT

1

What aspects of a
relationship should
we aim at with a
Companion, in terms
of such conventional
categories as
emotion, politeness,
and affection?

2

Even if it is not a
robot, in the sense of
a free-moving entity, should it
have a screen, and should it
have a visible avatar for
communication, whether human,
animal, or abstract?

3

Need the Companion have
one identifiable personality,
or several, and should the
user be able to choose the
Companion's personality or
shift between them if there are
several, making a Companion
lively and upbeat one day,
and subservient the next?

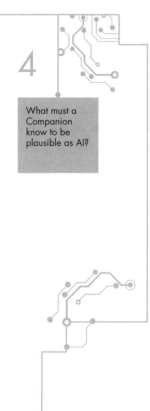

4

What must a Companion know to be plausible as AI?

5

What safeguards are there for the information content of such a Companion, in the sense of controlling access to its contents for the state or a company, and how should a user best provide for its disposal in case of their own death or incapacity?

Let us look at the first four issues in turn. The fifth will be considered in a later chapter.

> EMOTION, POLITENESS, AND AFFECTION

British psychologists Christine Cheepen and James Monaghan presented results some years ago that users of automata such as ATMs are repelled by excessive politeness and endless repetitions of "thank you for using our service," because they know they are dealing with a machine and such feigned sincerity feels wrong. This suggests that politeness is very much a matter of judgment in certain situations, just as it is with humans.

We know, since the original work of American psychologists Clifford Nass and Byron Reeves in the 1990s, that people display some level of feeling for the simplest machines, even PCs—in their original experiments, people avoided criticizing the performance of their own PCs if they could! However, the focus is on human psychology when faced with entities known to be artificial and does not bear directly on the issue of whether Companions should attempt to detect emotion in what they hear from us, or attempt to generate it in what they say back.

The AI area of "emotion and machines" is somewhat confused and contradictory. It has established itself as more than an eccentric minority taste, but as yet has little concrete to show beyond procedures for detecting "sentiment" in incoming text and—even though such software has been in great commercial demand—rests on little more than detecting the presence of certain emotionally "loaded" words. This strand of work began as "content analysis" at Harvard University's psychology department many decades ago. It showed that while prose texts may offer enough length to allow the assessment of a measure of sentiment, that is not so easy with short dialogue interactions. The technology rested almost entirely on the supposed sentiment value of individual words, which ignores the fact that their value is so dependent on context. "Cancer" may be marked as a negative word, but the utterance "I have found a cure for cancer" is presumably positive and detecting the appropriate response rests on the ability to extract information way beyond single terms. Not being able to do that has led to many of the classic absurdities of chatbots such as congratulating people on the death of their relatives.

At deeper levels, there are conflicting theories of emotion for automata, not all of which are consistent, and which apply only in certain kinds of discourse. So, for example, the classic theory of American psychologists Stacy Marsella and Jonathan Gratch that emotion is a response to the failure and success of our plans covers only those situations that are clearly driven by plans. But companionship dialogue is not always closely related to plans and tasks; people like just to chat much of the time without wanting anything done, and that may well apply to Companion machines too.

All this makes many emotion theories look primitive in terms of developments in AI and NLP elsewhere. British philosopher John Wisdom once said of philosophical discoveries that they are often the "running of a platitude up a flagpole," and theories of emotion have something of that quality. But, without doubt, there has been progress in incorporating emotion into artificial devices.

An important issue to be settled in a Companion's design is whether it should invariably try to cheer a user up if miserable, or rather track them by being sad with a sad user and happy with a happy one? There is no general answer to this question and, indeed, in an ideal Companion, the preferred approach would itself be a conversation topic—for example: "Do you want me to cheer you up or would you rather stay miserable?"

← Sophia, a humanoid robot, addresses the audience at a 2017 tech conference in Russia.

> ## WHAT SHOULD A COMPANION LOOK LIKE?

I prefer a faceless Companion—the furry purse, warm and light to carry, chatty but with full Internet access and probably no screen. If that feels too impersonal, such a Companion could always take control of a nearby screen or a phone if it needed to show itself. If there is to be a face, the question of the "uncanny valley effect" always comes up. This is a phrase coined by Japanese roboticist Masahiro Mori, who argued that people get increasingly uneasy the more something artificial is like them. I do not feel this, and certainly not with an avatar so good that I could not be sure initially it was artificial. "Emily" from Manchester was an early demonstration from a British company. She is ten years old now but seemed utterly real, and not uncanny at all, until at the end of the video when she takes her own face off to show she is artificial!

> ## ONE COMPANION PERSONALITY OR SEVERAL?

Some have argued that having a consistent personality should be part of being a Companion, but you could argue that, although this is true of a human companion— multiple personalities being a classic psychosis—there is no reason why I should want just one type of personality in a Companion. Perhaps a Companion should have a personality adapted to its particular relationship to a user at a given moment— such as when the user wants a Companion to function as a strict gym trainer. It might be possible to give a user control over a range of Companion personalities—something you could think of as an "agency" of Companions, rather than a single "agent," all of which shared access to the same state and history of the user.

> WHAT MUST A COMPANION KNOW?

Dogs make excellent companions and know nothing, in some sense of "know." Colby's PARRY program, which we met earlier (see pages 90–93), was the best computer conversationalist of its day (the 1970s) and possibly since, and famously "knew" nothing. On the other hand, it is hard to relate over a long term to an interlocutor who has no memory of what it, or you, have said in the past. It is hard to attribute much personality to an entity with no memory and little or no knowledge.

Much of what a Companion knows about its owner it should elicit in conversation. Yet, much could also be gained from publicly available sources, such as using Facebook to find out who its user's friends are. Current information extraction technology that I described in Chapter 5 allows a reasonable job to be made of going to Wikipedia for general information. When, say, a world city is mentioned, the Companion can then glean something about that city and ask a relevant question such as:

"Did you see the Eiffel Tower when you were in Paris?"

This again gives a plausible illusion of general knowledge.

John McCarthy always maintained that the real challenge for AI was not having exotic or detailed knowledge but common-sense knowledge—what exists below our levels of consciousness, such as that dropped things fall, and fingers go into water when pushed but not into tables. Some of this can be coded in the inference rules a Companion will need, such as that sisters share parents, but much of it is below the level of straightforward rules, which is exactly what led Hubert Dreyfus and others to argue that plausible AI would need the ability to learn as we do by growing up, rather than by using existing forms of machine learning or hand-coding. However, the great improvements in such learning in recent years—as we saw in examples from speech recognition to machine translation in Chapter 5—suggests that the jury is still out on this, even if the methods that have proved successful in computers may not be those that humans themselves use.

THE POWER OF PETS

Not all successful emotionally aware Companions have language. One of the most attractive of all current artificial pets is the Japanese PARO seal. It is furry, has no language, but wiggles like a real animal or baby. PARO's secret is its many servomotors under the skin that give it an animallike feel and let it give pushback when held. The notion of pushback or feedback is an old one, going back to nineteenth-century physics. It became the central cybernetic idea, and in this case the basis for American mathematician Norbert Wiener's notion that activities such as walking are only possible because of constant information feedback to the brain from our "servo" muscles in contact with the ground. The French "Nabaztag" rabbit toy was designed so that two different owners, usually far apart, could express their feelings via the Nabaztag held by the other, which glowed in a number of colors to indicate the feelings of the sender (such as blue for "sad").

The importance of these Companion toys is that they make emotion rather than language central to communication and relationships. Everyone knows that in relationships with pets—a central relationship for many—this is the case. Strong emotions are aroused by actions such as stroking, but there is no verbal content or feedback.

Nabaztag was an early media device for lovers separated by long distance: each had their own Nabaztag, and it received and sent messages, but also showed emotion by color and ear movement.

In 2020 Boston Dynamics released their robotic "dog" Spot for commercial sale, giving businesses the means to conduct complex and dangerous tasks more safely.

ANOTHER COMPANION PARADIGM: THE VICTORIAN COMPANION

The first fictional Artificial Companion is, of course, Frankenstein's monster in the nineteenth century. Mary Shelley's creature was dripping with emotions, and much concerned with its own social life:

> **"Shall each man," cried he, "find a wife for his bosom, and each beast have his mate, and I be alone? I had feelings of affection, and they were requited by detestation and scorn. Man! you may hate; but beware! your hours will pass in dread and misery, and soon the bolt will fall which must ravish from you your happiness forever."**

This is clearly not quite the performance that any modern Artificial Companion would be aiming at but, before just dismissing it as an "early failed experiment," we should take seriously the possibility that things may turn out differently from what we expect, and Companions, however effective, may be less loved and less lovable than we might wish. Some have argued that we must find out what kinds of relationship people will want with Companions, as opposed to being technologists and just deciding and then building what they believe people want.

Let us remind ourselves for a moment of
another sense of the word "companion"
as applied to the role of the Victorian lady's
companion. Forms of this still exist, as in
the Web posting:

COMPANION JOB

**posted: October 5, 2007,
01:11 AM**

**I am a 47-year-old lady
seeking a position as
Companion to the elderly,
willing to work as per your
requirements. I have been
doing this work for the past
11 years—very reliable
and respectful.**

Location: New Jersey

Salary/Wage: Will discuss

Education: College

Status: Full-time

Shift: Days and Nights

This kind of offer has become more closely identified with the social services than it would have been in Victorian times, where the emphasis was on company, preferably educated company, and diversion, rather than care. But you could nevertheless, and in no scientific manner, risk a listing of features of the ideal Victorian companion:

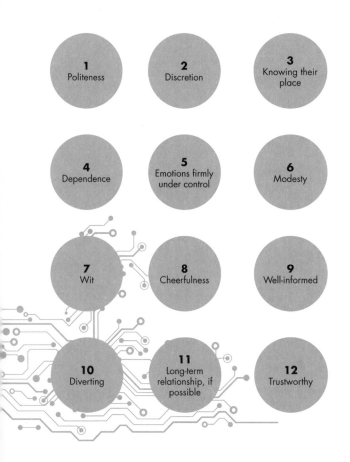

1
Politeness

2
Discretion

3
Knowing their place

4
Dependence

5
Emotions firmly under control

6
Modesty

7
Wit

8
Cheerfulness

9
Well-informed

10
Diverting

11
Long-term relationship, if possible

12
Trustworthy

↗

HAL 9000 was the onboard computer in a spaceship in the 1968 movie *2001: A Space Odyssey*, directed by Stanley Kubrick. It had its own goals, and the astronauts could not control it.

We might add that looks are irrelevant, and, to define the role further, state that limited socialization between Companions is permitted off duty—a point to which we will return shortly.

The emphasis in the list is on what the self-presentation and self-image of a tolerable Companion should be—crucially, that overt emotion may not be wanted at all. It is important to say this because many research Companions now being produced press their explicit "emotion" on an audience all the time, and I think this may be a mistake.

On the other hand, the pet–Companion analogy suggests that overt demonstrations of emotion are sometimes desirable and sought by pet owners, especially from dogs. Language, however, disguises emotion as much as it reveals it, and its ability to please, soothe, and cause offence are tightly coupled with linguistic expertise—as opposed to the display of gestures and facial expressions. We all know this from non-native speakers of our languages who frequently offend, even though they have no desire to do so, and often have no awareness of the offence they cause.

What name to call someone by, or whether to use titles such as "Sir," "Mister," "Miss," or "Missus," are complex matters, now made harder with the rise of new gender pronouns. All this is known intuitively by native speakers but not by outsiders, who are unlikely to have received a thorough grounding in such matters. When programming a Companion, this presents

exactly the kind of problem McCarthy identified as the unconscious, intuitive knowledge AI finds so hard to code. I personally find the "lady's companion" list opposite an attractive one: it avoids emotion beyond the linguistic, it implies care for the mental and emotional state of the user, and I would personally find it hard to abuse any computer with the characteristics listed.

Many potential interactive situations are, at present, wildly speculative, such as a Companion acting as its owner's agent, on the phone or Web, perhaps holding power of attorney—where you can act for an incapacitated person—or, with the owner's advance permission, perhaps even being a source of conversational comfort for relatives after the owner's death. Companions may not all be nice or even friendly. To stop us falling asleep while driving they may tell us jokes but will probably also shout at us and make us do stretching exercises. China is already creating a range of bad virtual boyfriends, said to be causing considerable social problems. Long-voyage Companions in space will be indispensable cognitive prostheses for running a huge interplanetary vessel and its experiments. Hollywood already knows all that, creating, in 1968, the terrifying bad Companion HAL 9000 in the movie *2001: A Space Odyssey*. Most of these situations are at present far off—but perhaps we should be ready for them.

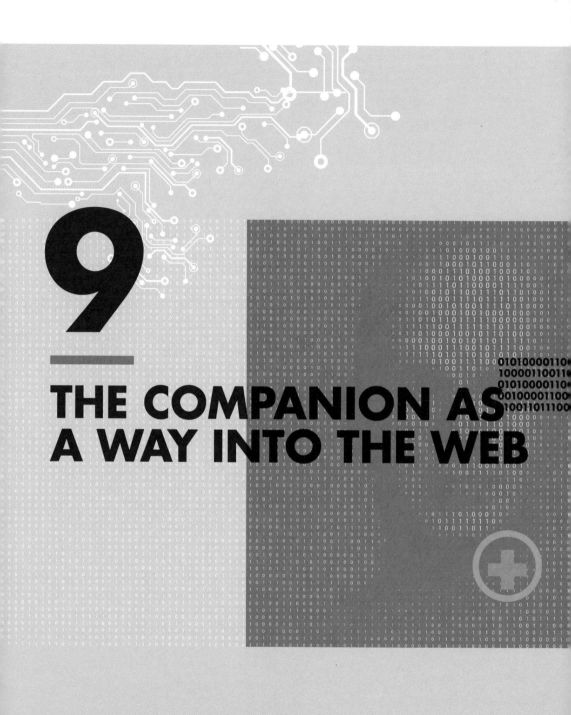

9
**THE COMPANION AS
A WAY INTO THE WEB**

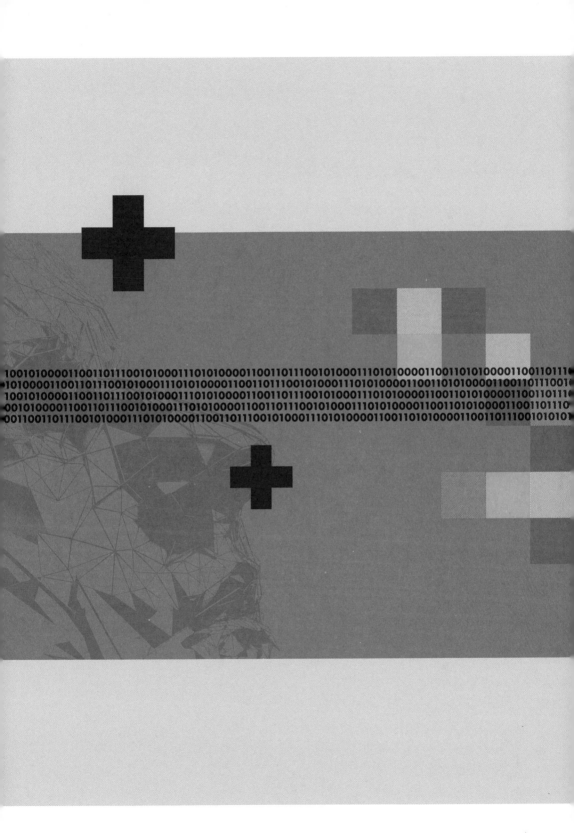

As late as 1996, the American newspaper *Investor's Business Daily* quoted a senior market intelligence executive as saying:

"... the Internet and World Wide Web phenomenon will go from an intoxication stage to a hangover stage during the next two years."

A few years later *The Guardian* newspaper reported a similarly misguided complainant who said: "the Internet is killing their trade because customers ... seem to prefer an electronic serf with limitless memory and no conversation." It is this lack of conversation with the Internet that the Companion might remedy. I assume in what follows that any such Companion, such as the Senior Companion of the previous chapter, can be seen as a prime exemplar and incarnation of all major AI technologies brought together in one place, rather as the German composer Richard Wagner called opera the *Gesamtkunstwerk*, the "total work of art."

I introduced Companions in the last chapter as specialized computer agents that could converse and assist groups of people such as the elderly. That assistance will almost certainly go way beyond helping organize their lives and memories, and will extend to interacting with the electronic world outside for them, as Web agents. Web agents already make deals and transactions of all sorts with each other, and learn to trust each other, in, for example, the world of banking, where agents in their thousands clinch money transactions between financial institutions.

These activities may come to require exactly the kind of computer agency a Companion will be able to offer to a person when dealing with the Web as it becomes more complex. To put this very simply: the Web may become unusable for nonexperts unless we have Companion-like agents to manage its complexity for us. One reason for this is that there is now just too much data to cope with in our lives, seen with external data we need to access as well as all the data we generate ourselves (such as thousands of photographs of ourselves and our meals). Another reason for possible Companion Web help is that people are not very good at online searches—but do not know this. They can input names and sex and holidays into Google and get something useful back, but hardly anyone ever looks beyond the first page of what a search engine returns, and they rarely

HOW CHATBOTS WORK

↓

A chatbot is a simple conversational agent
that uses little AI. It is widely employed by
companies as part of their Web front-ends
to deal with customers without human aid.

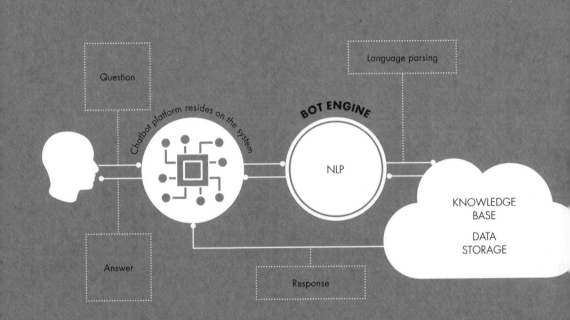

filter their search results—keeping in certain terms and dropping others—in ways that Web searches allow but which nonspecialists rarely know about.

So, the Web—in the sense of the totality of our electronic personal information and intellectual activities in cyberspace— may have to develop more humanlike characteristics at its human interface to survive as a usable resource and technology. Simply locating a particular individual on the Web, when most of the world's population have a Web presence, will become far more

difficult and time-consuming than it is now. Companions will then be needed by everyone, not simply the old, the young, and people with disabilities. It may become impossible to conceive of the Web without some kind of human face that renders it personal, which is a short step from the kind of personalization that has come to mean "getting only the news you agree with"— which has in turn become part of the problem. You could say that AI and Web technologies have created these problems, and they are going to have to solve them.

DECEPTION AND IDENTITY

Newspaper readers are all familiar with stories of chatroom users who pretend to be what they are not. Sometimes this is a quite innocent pretense—little more than hiding behind a pseudonym during conversations and sometimes pretending to be a different kind of person. This is barely different from legitimately checking into a hotel under an assumed name. The Victorian game of sex-pretending, on which Alan Turing based his famous imitation game for computer intelligence (see page 18), has come back as a widely played reality. The problems—and they are very real—arise when impressionable people, usually children, are lured into meetings with people they have encountered under false pretenses.

Such issues as real people on the Web pretending to be someone else are closely related to a wider problem of *artificial people*, or "bots," pretending to be real, possibly to gather information or to persuade others of political views or how to vote. Claims appear regularly in the media that large numbers of Facebook user accounts are actually bots, and many of these will have been created by the Western national security agencies to infiltrate discussion groups seeking to chat to people who have what they consider dangerous political views. That is a well-funded area of NLP technology, which also works out what the influence relationships within such a group are, and who, if anyone, is controlling the spread of ideas in it. Our newspapers are keen to find examples of Russian propaganda bots spreading false beliefs about current political issues, while failing to report the vast sums our own governments spend on doing just that here and in other countries. But this need not all be bad news. Companion agents could offer the possibility of managing our multiple identities for us—should we choose to have them—using different identities (legally) in differing Web transactions.

HOW A WEBBOT OPERATES

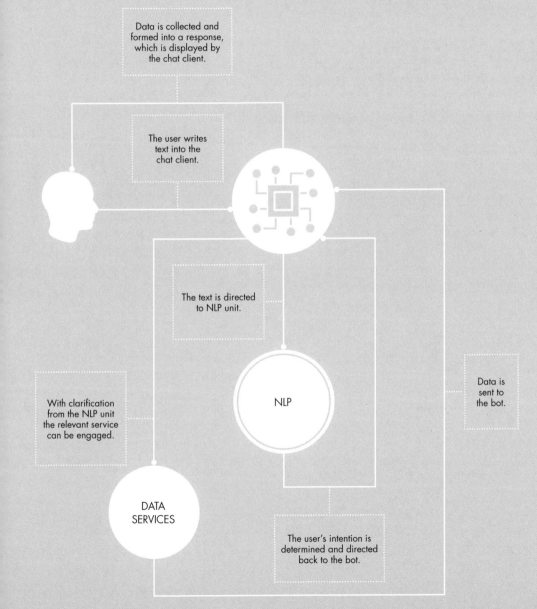

A webbot is usually a chatbot with access to a victim's personal data and which appears to be a human agent on social media platforms such as Facebook.

Data is collected and formed into a response, which is displayed by the chat client.

The user writes text into the chat client.

The text is directed to NLP unit.

With clarification from the NLP unit the relevant service can be engaged.

NLP

Data is sent to the bot.

DATA SERVICES

The user's intention is determined and directed back to the bot.

DISCRETION IN A COMPANION

I discussed in Chapter 8 the nineteenth-century idea of a Victorian lady's companion. I noted that one of her key features was discretion—something it may be worth thinking a little more about in today's terms. How much might we want to keep secret from a Companion if we did not trust its discretion, or its commitment to just us as the owner? The household presence of Alexa is already causing such concerns, with users becoming aware that Alexa is listening to domestic conversations and almost certainly reporting them centrally to its maker, ostensibly for "learning and improvement" purposes.

More generally, will future Companions share their knowledge behind the backs of their owners, in the way that Victorian domestic servants were generally believed to do, but which ladies' companions were supposed not to? How confident will you feel, in the middle of a family feud, that the instruction:"Don't tell Billy this" to an Artificial Companion will always be obeyed? Much of this will be debatable because Companions will normally deal only with one person— which is what makes their speech recognition problem so much easier. They will be trained for a single speaker, except when, say, making phone calls to an official, a friend, or a restaurant. In 2018, Google demonstrated an automated call service that could reserve a restaurant table—although this has been possible for more than twenty years now. Like much AI, the gestation period to market is much slower than many believe.

The notion of a stored fact that must not be disclosed is simple to code, but that same fact must, to preserve the secret, not take part in inference processes either. If it is a secret that Tom is really a Russian, then the Companion should not do inferences such as [IF X is of

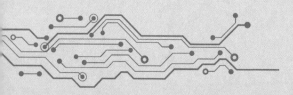

nationality Y, THEN X will normally speak Y] and suddenly come out with, "I assumed Tom could speak Russian," which would give the game away.

On the other hand, a Companion gossiping behind an owner's back could be a positive development that we might want to encourage. Imagine the elderly person in a nursing home, too shy to approach a fellow resident for a lunch together. This would be something best settled between their Companions, each knowing the tastes and habits of their owner, to whom the "date" could be presented as a fait accompli. Again, many Companion-to-Companion interactions will be between an individual's Companion and some form of "public Companion," such as one that takes restaurant bookings based on a user's tastes; or at a hospital, where a hospital Companion would triage incoming patients who may not be articulate about their condition, based on detailed knowledge of the user's medical records revealed by their Companion. When traveling, this Companion-to-Companion

interaction in, say, a hospital could also combine with translation where the respective Companions worked out how to communicate across a language barrier.

In all these cases, Companion-to-Companion communication could be of obvious benefit to a user even if confidential information was at risk of disclosure. The user might have said: "Never tell anyone I'm HIV positive," but in the hospital environment that constraint should obviously be overridden and the user's condition revealed. You could say that secrets may be relative to a situation and that there may be nothing more complex in a Companion's guardianship of secrets than there is in explicit restrictions you could give to human hearers. In some situations, a Companion might owe a public loyalty rather than one to an owner. Consider Companions provided to long-term prisoners for their mental health. Long years of conversation might well inadvertently reveal criminal associates' names, or even confessions, to the authorities.

SECURING PERSONAL DATA

A Companion that had learned intimate details of a user's life over months or years would certainly have contents needing protection, and many interests—commercial, security, governmental, research—might well want access to them, or even to those of all the Companions in a society. If a society moves to a legal state where your personal data is your own, with the owner or originator having rights over sale and distribution of their data (which is not at all the case now in most countries), then the issue of the personal data elicited by a Companion would automatically be covered.

A Companion would clearly be useful to the police when they want to know as much as possible about a murder suspect. So, there might be an issue of whether talking to a Companion constituted any kind of self-incrimination—and in some countries that form of communication might in future be protected. Some might want a relationship to a Companion put on a similar basis to a relationship to a priest or doctor, or even to

DATA AFTER DEATH

There are already four major types of "death site" on the Internet that can hold postmortem data:

- memorial and tribute sites created for the already dead;
- "locked boxes" of assets and secrets for survivors that protect the individual's interests after death;
- "legacy" sites containing final wishes and emails to be revealed or sent after an individual's death;
- "life story" sites that manage autobiographical material for an individual creator, to leave some form of self-presentation of their life.

MIXED MESSAGES

In the UK, there is at present a contradiction between the forces of two parliamentary acts. On the one hand, you are not allowed to publish, or even store, the names of, say, the members of your sports club on the Web or a computer without registering your intention. Nor can you legally publish on a noticeboard the names of your students or their grades achieved. You may not even be able to publish a remark about someone's health, lest that imply access to privileged health information. There is an obvious contradiction between those constraints and the freedom of access that companies and the State have to electronic transactions of all kinds. It is common knowledge that British and American state security process a substantial percentage of their citizens' emails and phone records as and when they choose and without any need of access to the courts, even though a journalist has no such protected privilege. A recent editorial in *Vanity Fair* succinctly stated that governments seemed to want "transparency for you, but privacy for me."

a spouse, who cannot always be forced to give evidence in countries such as the US and UK.

More realistically, a user might well want to protect parts of their Companion's information from particular individuals—to have discretion, in the term I used above—such as: "This must never be told to my children, even when I am gone." It is not hard to imagine a Companion deciding to whom it may divulge certain things, selecting between classes of offspring, relations, friends, colleagues, and so on. There will almost certainly need to be a new set of laws covering the ownership, inheritance, and destruction of Companion–objects in the future. A Companion would probably be designed as a gleaner of data from open sites such as Facebook and from the user's own holdings of documents and images. But,

since some Companion applications will be health-related, there is no reason why this user-based data hoard should not ultimately incorporate health information. In countries such as Spain, where everyone has access to, and ownership of, their own health information, you would expect that. This is opposite to the focus of much current public discussion, particularly in the US and UK, where public and private organizations are almost certainly more intrusive than elsewhere, and where, consequently, much public opinion is either negative or ambivalent on these issues.

IDENTITY AND IDENTITY SHIELDING

If a Companion is to be an interface to the Internet, say, for a user who is technologically unskilled yet who must conform to the standards of identity that society requires and imposes, then that Companion will have to understand identity to some degree and possibly be able to manipulate slightly

↓

The World Economic Forum has identified six components of digital identity; of these, authorization addresses the thorny issue of proving (and protecting) personal identity online.

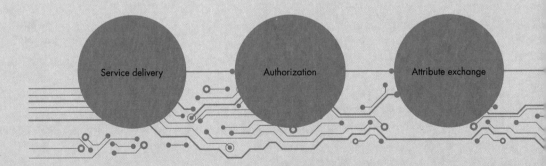

Service delivery

Authorization

Attribute exchange

different forms of it. In the US and UK, identity is currently established by a range of items with numbers, from passports through to credit cards, driver's license, and tax numbers (some with associated passwords), with the Social Security number having a definite primacy in the US. In most EU countries, there is a single ID number, of which the simplest is the lifelong single Personnummer in Sweden. States prefer a citizen to be identified by a single number, and in the UK there is currently strong pressure for something closer to the Swedish model, although UK law has, now, no clear definition of identity, with legally registered unique names and addresses, as in most of the EU. There is no legal problem in the UK with having several identities simultaneously and bank accounts for each, so long as there is no intention to defraud.

This is important since identity checks are the basis of all Web transactions. If a Companion is to deal with a person's affairs it will need something approaching a "power of attorney" or at least an understanding of how identity is established in Web transactions. It will also require a method for establishing that its owner approves of what it is doing in individual transactions, in case of later disputes—for example, from angry relatives if an elderly person's money has been spent.

A more liberal data regime would be based on identity shielding, whereby, if our data is open to the state and to corporations, so should it be for individuals to use as they choose. The key principle must be an

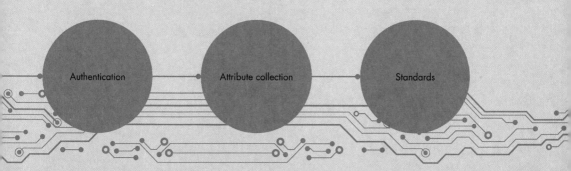

Authentication

Attribute collection

Standards

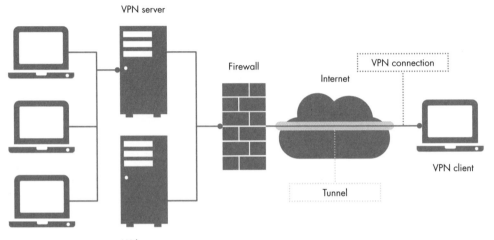

VPN server

Firewall

VPN connection

Internet

Tunnel

VPN client

Web server

↑

A VPN is a way of connecting to the Web, bypassing the firewall barriers set up by companies or governments to keep you out, because you are in a different country, for example.

individual's control of their data for their own purposes, such as their right to know what is held on them, way beyond current minimal access that exists to credit bureaus and their records. British Web guru David Birch has expressed a clear view on how an individual could combat state and corporate identification of individuals by a system of controlled and limited revelation of identity in transactions.

Abbie Hoffman, of the 1960s Yippie movement in the US—one of the first great counterculture movements of modern times—was already sensitive to these issues, and his advice then (to Americans) was to give a different Social Security number every time one was requested. But that trick, though fun, cannot survive in the age of giant data processing and instant access to identity banks based on such numbers. Birch's suggestions are complex, but their essence is that we should encourage systems where we reveal as little or as much of our identity as we choose at each transaction, so that no giant global data footprint outside our control can be assembled, whether by state or corporate actors. Such a concept is perfectly consistent with the UK and US legal concept of identity, or rather the lack of one, where you may still legally use any name you like.

The difficulties with the general principle of "identification minimization" are in the details and, particularly, how we minimize the inevitable loss of information for criminal detection and national security. There are many technical devices that could be more widely used than they are, including ways to shield our computers from data gathering and linking of data.

Almost certainly, there will be increasing commercial provision of methods for avoiding the identification of a machine's address, geographical location, and so on. This used to be associated with anonymous remailing sites and the anonymity of activities such as pedophile rings and terrorist groups. But such facilities will become increasingly available; American Internet security expert Lance Cottrell argued a decade ago, and correctly, that disguising your identity by such methods to operate "anonymously" on the Internet

PROTECTING IDENTITY

A simple identity limitation could be used on various states' motor vehicle licensing departments' websites. It can check, say, insurance and any car test details, and issue a record that it stores and which police and insurance companies can access. But there is no reason why anyone should have to reveal the name of a car's owner in that process, any more than with the use of an anonymous prepaid phone. The police have many ways of finding out who is driving a car on a given occasion, and ownership is no guide to that for any given accident or other event. Cars are full of biomarkers that could well be sufficient for investigative purposes.

would be a growth industry in the coming years—for all the reasons given here. Surveys show that children may already be adapting to this possibility by creating alternative identities on social media. Facebook decided in 2019 to enforce a previously little used rule of "no anonymity on Facebook," even though there has been no way, in general, of detecting such anonymous membership. It is probably revealing that Facebook has raised this issue in connection with its own commercial functioning in China, where the government is keen to block anonymous criticism and wants to root out such anonymous critics with Facebook's help.

WEB "ANONYMITY" OVEREMPHASIZES THE IMPORTANCE OF NAMES

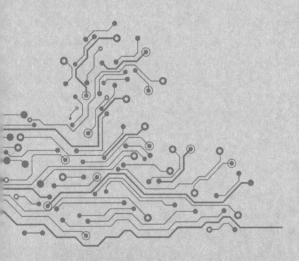

Tim Berners-Lee has said that soon anonymity in computer transactions will be impossible because of the ancillary information available—based on location, past search actions, and so on—from which our identity can be inferred even if we shield our name explicitly. David Birch has highlighted some of the mechanisms necessary for avoiding this (should we wish to), such as payment mechanisms that are Web-usable but as anonymous and as reliable as cash. Bitcoin was designed to offer this but is enormously complex and expensive in terms of computation—the wider use of payments through anonymous digital phones would be an obvious and cheaper start. There are many technical details and possibilities, which could also involve the use of trusted agents, such as Companions, to act for us at commercial and government portals— trusted, that is, by both the user and the provider. What is needed is the will to move to such a system if enough of the population find the current situation (and its likely developments) intolerable.

However, we must not be led into overemphasizing the importance of locating a unique name to identify an individual, and the very word "anonymity"—having no name—leads us into this trap. Berners-Lee's remark above does not emphasize the issue

of names, and this is a crucial point, one that serves to mitigate some of what has been argued for so far. A citizenship roster for the citizens of a state, each having an identification—a personal number, say— is not the same as a link to a unique name, nor should it be. The French state has traditionally demanded the address of every citizen and that their name be drawn from a roster of acceptable names. English-speaking societies, however, manage without either of those conventions. Your name is not crucial to any of this identification, since only biomarkers are fully reliable for that, especially since very few names are "Google-unique," and their degree of ambiguity varies from society to society depending on social or historical factors. Sweden and Arabic-speaking countries, for example, have relatively small name lists and high name ambiguity. The resolution of such ambiguity is now an NLP technology for settling issues such as "Which George Bush is this document about?"

Names, addresses, and dates of birth are only some among many features of use in locating an individual—down to an index number, to a current spatial location, or to a unique biomarker. They are not special in the way we may think they are. Having a (correct) name may give us no special "hold" on an individual if we have enough other features, as is known to students of movies and television from *Mad Men* to *The Return of Martin Guerre*, where the message is: "This is you, no matter what you call yourself." Everything short of that is the same situation we encountered in Chapter 3 when discussing meaning in the Semantic Web—symbols that make up URLs only lead in the end to files that are more symbols; they never reach actual hard-bound books.

Perhaps the truth will be that Internet tracking and features give a "cloud" of probabilities as to who an individual really is, and this proves sufficient for many commercial (if not all security) purposes. In this cloud, the individual's name is only one feature and perhaps not the most important one.

This may seem an unduly technical analysis of the issues, but I believe it remains to be shown that there is any distinctively ethical component to the seeking, finding, giving, or withholding of an identity, at least in the sense of a name, as seems to have been generally assumed in recent cases such as that of the undercover policemen who conducted affairs under assumed names with those they were spying on.

"Ownership" of an identity, personal data, and so on clearly has an ethical dimension, at least for those who believe that their ethical sense is tied to their sense of identity and that it may not be possible to have one without the other. I would prefer to see the issue in terms of providing legal incentives for Internet service providers to give identity protection to those who want it, such as by the badging of sites that *require less identification* to carry out transactions.

Ethical questions hover widely in these issues—for example, many Western commentators seem to accept without question that the search engine Baidu in China is simply an agent of the state. But the reactions of PayPal, Amazon, and other companies in the US to WikiLeaks—the group that published national security information on the Web—by shutting down its financial and communication activities despite its principal not being convicted of any crime, is very much a scenario of Western companies acting like state agencies. US President Dwight D. Eisenhower famously warned the US, at the moment of his retirement, of the danger of the emerging military-industrial complex; we might now argue that the threat is threefold: the military-industrial-information complex.

In 2018, Tim Berners-Lee expressed his disillusion with the hate- and disinformation-laden Web that his creation had turned into and gave suggestions for personal control of data of the sort I have been discussing. These he presented under the name of a new project called Solid. It may be that human nature is such that the distancing from face-to-face communication that the Web provides unleashes emotions and immoderate behaviors that no technical constraints can ever remedy, and humans will simply have to learn to change their instinctive behavior. On the issue of truth versus falsehood on the Web, the prospects may be brighter as NLP begins to get a grip on the issue of detecting and identifying more reliable information sources from less reliable information ones, and I shall turn to this in the next chapter.

→

Tim Berners-Lee's vision for a safer, decentralized Web involves Solid Pods (above) to store and protect individuals' data, and Solid servers to connect multiple Pods.

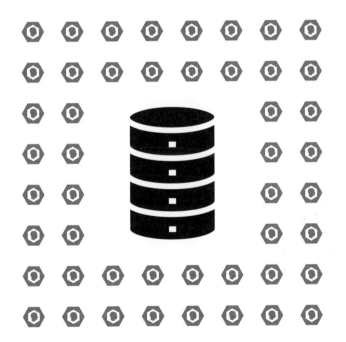

10

WAR, ETHICS, AUTOMATION, AND RELIGION

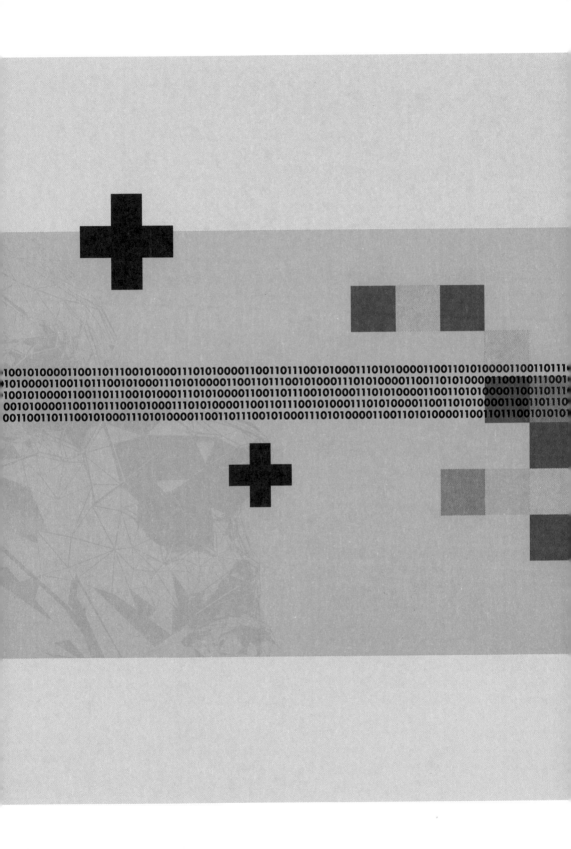

In 2017, a computer beat the world's number-one player of Go—a much harder game to play or automate than chess. It secured victory with a novel, creative move, which, at the time, the master simply could not understand. A year later, an automatically produced work of art sold at auction in New York City for nearly half a million dollars. As someone must have said by now: AI changes everything.

In this chapter, I want to touch briefly on some philosophical and political issues that AI has affected or created. It must be borne in mind that these express only the author's views because there is no agreed consensus on them, as there often is on parts of the technology itself. The issues range from ethical examples to machine consciousness—and beyond these issues, it is worth noting the sheer level of dematerialization of our world that AI and its associated technologies are bringing about. Canadian-American linguist and futurologist Steven Pinker has called this "peak stuff," and noted that the average Briton in the first decade of this century went from consuming 15 metric tons of "stuff" a year to 10. This, he said, was closely connected to the fact that his intelligent phone had freed him from a telephone, a watch, a tape recorder, a radio, an alarm, a calculator, a dictionary, an encyclopedia, an answering machine, a phone book, a camera, a camcorder, a Rolodex, a calendar, road maps, a flashlight, a fax, a compass, an outdoor thermometer, a metronome, and a spirit level.

AUTOMATION AND THE LARGE-SCALE LOSS OF JOBS CAUSED BY AI

There is a pessimistic tradition of technological determinism for society, going back to the easy lives of the Eloi in H. G. Wells's novel *The Time Machine*, which are contrasted with those of the dark, subterranean, post-proletarian Morlocks. AI and the Internet serve the academic and professional groups who invented and developed them very well, but it is not yet clear that they yet serve most of the population as well—beyond cheap vacations and pornography—although the pandemic of recent years has probably hugely increased Web use from home.

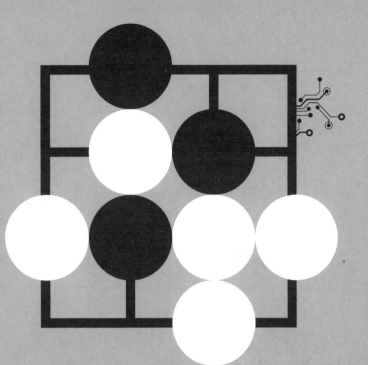

→ The Go game is ancient and more complex than chess. DeepMind's player has now beaten the Go world champion.

Many prognosticators claim that up to half of all current jobs will be automated by AI within a few decades—and not just those of truck drivers but also doctors and journalists. The labor component in car production, for example, has fallen from 70 percent to 48 percent in a decade. But what *new* jobs will be created, and can all those displaced become state employees with titles like "compliance officer"? World Economic Forum reports continue to argue that more jobs will be created than are destroyed by automation, and a glance at standard UN figures shows that the countries with the highest level of robot installations, such as Japan, also have the lowest unemployment.

So there is much evidence to be found on the optimistic side, on top of the simple historical fact that all previous phases of automation have been followed by many new jobs. Our ancestors could not have imagined the nonagricultural future most of us enjoy, free of harvests and backbreaking work. There also seems to be a general understanding that if profits continue to accrue mostly to the owners of AI capital—machines and software—much of that will have to be recycled to the rest of the population, employed or not, if developed societies are to remain stable. The Universal Basic Income is one suggestion for doing that. On the other hand, total hours worked have fallen in developed societies, and the alternative possibility that the nonworking population will continue to grow cannot be ruled out.

ETHICAL ISSUES ARISING FROM AI

Discussions of ethics in relation to AI, and the moral choices artificial systems (such as autonomous cars) may have to make, has been dominated by what is called the "trolley problem." This is an idea from philosophy teaching in the 1970s in which a railroad controller must decide whether a runaway train should be left to kill several people or should have its track switched to kill one otherwise safe person. A long discussion in academic journals has led to no concrete results, and even philosophers are wearying of it.

Earlier, I noted how the rise of machine learning as the core AI paradigm has meant that we may not know exactly how our programs will make particular decisions in the future—hence the rise of research in XAI (explainable AI) and the current DARPA program to try to provide that (see page 124). The European Commission has legislated in a way that has been interpreted as demanding that deployed machine learning systems must explain their decisions, even though no one

yet knows how to provide what they seem to be requiring. The English political philosopher John Gray has argued that we—as human actors, that is, not as psychologists—have little or no insight into how we take decisions, moral or otherwise, and a part of modern psychology agrees with him. What would follow if machines and ourselves were in roughly the same position as regards the transparency of our ethical decision making?

Gray's starting point is that discussions in classical ethics have little or nothing to do with how humans or animals seem to act. He believes they act simply "like machines" (and he means that in a positive sense). For Gray, we do not calculate ethical rules or consequences before acting, as the ethics textbooks tend to assume (so neither should machines, he might have added). He may be right about the conscious processes of humans in action, but his position is also circular: humans do not act randomly, so there must be some causal explanation of what they do. We can barely imagine legal and social life

Philippa Foot's Trolley Dilemma was designed to teach ethical problems to students: should the signal be set to kill the one victim or the many?

A FATAL MORAL DILEMMA

An international survey, carried out by the MIT Media Lab and published in 2018, showed a striking difference between Asian and Western populations when asked if a runaway car should kill a baby or an elderly person—if it was destined to kill one or the other. By large margins, Asians voted to kill the baby and Westerners the senior citizen. This suggests automated cars may need different ethical software in different parts of the world.

IBM'S
FIVE KEY ELEMENTS
OF ETHICS

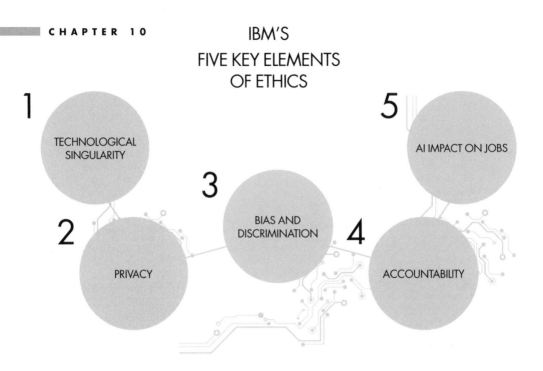

1 TECHNOLOGICAL SINGULARITY

5 AI IMPACT ON JOBS

3 BIAS AND DISCRIMINATION

2 PRIVACY

4 ACCOUNTABILITY

without this prop of *responsibility*, even if it is all or in part a fiction.

It is important to remember that traditional ethical thought, like AI reasoning itself, assumed such decisions to be matters of calculation from rules or working out the relative consequences of different actions. Ethical traditions require calculation, logical or arithmetical, as their basis, which is why they have appealed for so long to the computationally minded. "For reason ... is nothing but reckoning," as the English philosopher Thomas Hobbes said in the seventeenth century. But these ethical exercises are not real calculations that are ever carried out, and actual values are never assigned to possible outcomes in such discussions, even though, in the real world, automated systems such as cars have to make such decisions every day.

Swedish philosopher Nick Bostrom and American computer scientist Eliezer Yudkowsky have argued that machines must be programmed with comprehensible rules if we are to tolerate them among us, so that we can understand them and why they do what they do. Yet, if machines that take decisions are based on ML algorithms, it is not clear that such unambiguous transparency will be available, as we saw earlier. There will always be alternative explanations of any human behavior, too, and courtroom drama rests on that fact. In the case of machines, their situation may come to be seen as the same as our own unless something quite new is added alongside whatever it is they are programmed with; and saying that may be just another way of demanding XAI.

In Chapter 8, I discussed the notion of an Artificial Companion—a personal web agent

THOMAS HOBBES
1588–1679

Hobbes was a seventeenth-century English philosopher, indeed the first modern philosopher of international influence from England. He was probably the first to put forward the notion that became the "social contract," the idea that people give up some freedoms to a ruler or sovereign government so as to have peace and order. Hobbes was preoccupied that life in nature was "nasty, brutish, and short" and therefore some sacrifice of freedom was necessary to avoid this state of war "against all," as he called it.

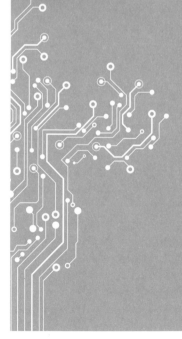

permanently associated with a human, gaining the maximum possible knowledge about its human "owner" via dialogue over an extended period of years, and which is designed to handle the vast quantity of personal and public data that increasingly we cannot. This Companion might also be able to supply the data from the past needed to make inferences about a person's basis of action, containing perhaps self-revelations (or confessions) by a Companion "owner" to the Companion that could be crucial for later ethical explanation. We can imagine a person, as a form of therapy, consulting their own ethical Companion to understand why they had acted as they had in the distant past.

Automated cars have focused people's minds on the issue of responsibility, and where blame may lie when a car or a Companion acts as a person's agent and something goes wrong. At present, Anglo-American law has no real notion of any responsible entity except a human, and to some extent corporations, at least if we exclude "Acts of God" in insurance policies. The only possible exception is dogs, which occupy a special place in English common law, where they have certain rights and attributions of character, separate from their owners. If you keep a tiger, you are totally responsible for whatever damage it does, because it is *ferae naturae*, a wild beast. Dogs, however, seem to occupy a middle ground as responsible agents, and an owner may not be responsible unless the dog is known to be of

XAI tries to map how a neural network makes its decisions and explain them so that functioning of the network remains transparent.

TRAINING DATA

"bad character." Such law might be a narrow window through which we might begin in the future to squeeze notions of responsible machine agency, different from that of the owners and manufacturers of machines.

It is easy to see the need for something like the dog example. Suppose a Companion told your grandmother that it was warm outside and, when she went out into the freezing backyard on this advice, she caught a chill and died. In such circumstances, you might well want to blame someone or something and would not be happy to be told that Companions could not accept blame and

HOW XAI WORKS

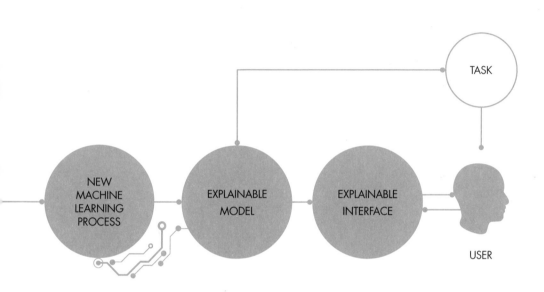

TASK

NEW MACHINE LEARNING PROCESS

EXPLAINABLE MODEL

EXPLAINABLE INTERFACE

USER

that, if you read the small print on the box, you would see that the company denied all responsibility and had got your grandmother to sign a document to that effect. Perhaps this would seem acceptable if the company then sent the Companion a small tweak so it never happened again.

This last story applies to nothing at present. Indeed, the Companion might point out, with reason, that it had read the outside temperature electronically and could show that it was a moderate reading and therefore the blame should fall on whoever was responsible for maintaining the thermometer.

What is obvious already is that Companions must be prepared to show exactly why they said the things they said and offered the advice they did, which is why the XAI project is so important.

THE ETHICAL QUESTION OF AUTOMATED WARFARE

Warfare, whatever its moral defects, has undoubtedly driven technology for millennia, and will continue to do so, we may safely assume. Major powers are currently spending vast amounts on AI, and some of our leaders now have a vision of military operations conducted entirely by automata—though there is still considerable investment in strengthening the human frame with exoskeletons, as discussed in Chapter 7, to create soldiers of enormous strength and speed. Fully automated armies need have no relation to the human form. They could be flocks of small boats or submersibles, small vehicles or drones, all equipped with cameras and powerful weapons.

The main advantages of automata are: the lack of human casualties on "our" side and, more interestingly, the overcoming of the antipathy of most soldiers toward killing. This last advantage has only emerged from recent studies, and it now seems that in the wars available for modern study more than 80 percent of casualties have been inflicted by only 10 percent of the soldiers, and most soldiers go to great lengths not to kill, and to hide that fact as well. Dave Grossman, who was both a US marine officer and a psychology professor, showed from detailed analysis of US Civil War rifles that many had been double-loaded so they could not fire, although the soldier appeared to his officers to be reloading.

Ethical arguments against particular weapons long predate AI. In the Middle Ages, for instance, it is widely believed that the Catholic Church decreed that crossbows were unfit for Christian use. The chief argument against AI-based weapons, urged by activists such as Irish computer scientist Noel Sharkey, is that AI technology is not sufficiently advanced to separate combatants from noncombatants as targets, and so war crimes will inevitably be committed if it is deployed. But this was always the case with weapons— and the bombings of Dresden, London, and Hiroshima in World War II show that on a vast scale. And it may be a temporary

phenomenon because, if automated cars can successfully separate cyclists from pedestrians—and they can—we may hope that battlefield discriminations may be possible in time, too, and be better than those made by soldiers. Again, much serious discussion here will come down to how truly autonomous such weapons are, and how far they act only with human guidance. Outside that battlefield, major powers are also spending extraordinary sums on cyberwar: bringing a country to its knees with intelligent sabotage of its infrastructure and without direct casualties.

→

The Red Army soldier statue in Tallinn, Estonia, at the heart of a possible cyber attack in 2007.

It now seems that in the wars available for modern study, more than 80 percent of casualties (the red crosses) have been inflicted by only 10 percent of the combatants (the red soldiers), and most soldiers go to great lengths not to kill, and to hide that fact as well.

AN ANTIPATHY TOWARD KILLING

Fully automated armies need have no relation to the human form, but could be flocks of small boats or submersibles, small vehicles, or drones in the air, all equipped with cameras and powerful weapons. The main advantages of automata are: first, the lack of human casualties on "our" side and, more interestingly, overcoming the antipathy of most soldiers toward killing, evidence for which has only emerged from recent studies.

RELIGION

There have been various tenuous links between AI and religion. Recently, cults have appeared in the US dedicated to the worship of AI artifacts, if and when they become in some sense "superintelligent." One early historical landmark is the long essay "God & Golem, Inc.," published by American mathematician and philosopher Norbert Wiener just after World War II. Wiener was the founder of cybernetics, the early form of AI. The essay is for the most part an assessment of the impact of the coming of intelligent machines on ethics and on religion itself. The Golem was a medieval Jewish concept of an artificial creature with humanlike properties, created by a Prague rabbi. At about the same time as Wiener wrote, both twentieth-century Austrian-British philosopher Karl Popper and British mathematician and cryptologist I. J. Good speculated on what it would be to have a machine that knew everything. French scientist Pierre-Simon Laplace in the eighteenth century postulated a "demon" that knew the position and velocity of every particle in the universe, something we now know not to be possible if quantum physics is correct. That level of knowledge of physics tends not to be what we think of as "knowing all the facts there are." However, the Web is certainly pushing toward "knowing everything," although, as I discussed in Chapter 3, it does not know what it knows in any real sense. In the Semantic Web (SW) project (see pages 52–53), AI is certainly working on that.

The relevance of this is that knowing everything, or omniscience, is what traditionally helps to define God in the Western tradition. AI can, therefore, be seen as moving into that territory. In most religious traditions with a Creator of the Earth, life, and everything, it is also assumed that the Creator is more or less beneficent. Traditions differ on this, and many find the Bible's Old Testament a bit morally unsound with all its endorsement of mass killing, although most assume that we, the created, are well disposed to our Creator. Note that this beneficence on the part of the

NORBERT WIENER
1894–1964

Norbert Wiener was an American scientist and mathematician who specialized in stochastic, or statistical, processes, and who did much to create the notion of cybernetics, a forerunner of modern AI. In cybernetics, the human is part of a universal system governed by the notion of feedback and learning. Wiener's notion of intelligence is therefore based on statistical probabilities and not on the kind of logical representations that dominated AI under the later influence of scientists such as John McCarthy and Marvin Minsky, though Minsky kept some allegiance to cybernetic ideas.

created is usually independent of any assessment of the Creator's intelligence—IQ is not normally thought of as a theological concept. In traditional Christian theology, God is omniscient and omnipotent (as well as good), with certain restrictions normally imposed by the laws of logic—which is to say, not so omnipotent that God can do two contradictory things. There is, of course, a gap between omniscience (knowing everything) and being intelligent—the intelligent do not necessarily know a lot, and those who know a great deal (such as all the digits of pi) are not always intelligent. But the two knowledge concepts do seem strongly connected.

All this has intellectual consequences for the question of whether an artificial "superintelligence," were one to emerge, would be well disposed toward us. There is certainly a fear among many—and Nick Bostrom has devoted a book called *Superintelligence* to arguing this very danger—that such an entity would be hostile. However, religious traditions seem to me to suggest strongly that creations (that is, us vis-à-vis a Creator or a superintelligence vis-à-vis us) are well disposed to those who made them. If that is true, we need not fear what we may create in the future with AI.

In futurist discourse, there is often a confluence between this strain and that of adapting or evolving humans (or some of them, anyway) so that they become immortal or superintelligent themselves. Groups or individuals who believe this are now called "transhumanists" and have close links back to eighteenth-century rationalism and humanism, moving on to the worship of humans, or "Supermen" to use the term coined by German philosopher Friedrich Nietzsche. Transhumanists and superintelligencers only differ in the source of the supreme intelligence, not the goal itself. This is not a new idea. One of Austrian composer Franz Schubert's most famous songs, "Mut!," is set to a poem by the nineteenth-century German writer Wilhelm Müller and contains the lines "Will kein Gott auf Erden sein/Sind wir selber Götter"—"If there is no God on Earth/Then we will ourselves be gods."

↓

Are the Web and AI
turning the human into a
different kind of life form?

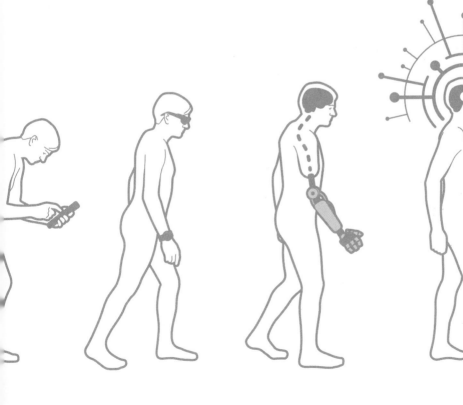

THE COMPANION'S LIFE AFTER DEATH

The disposal of memory, including secrets, by a Companion after a user's death is a complex subject, and we touched on "death sites" in the previous chapter (see page 178). This is undoubtedly an area with enormous commercial possibilities since the Internet makes actual death less apparent and immediate in the electronic world than in the real one. In the last week of his life, the eighteenth-century English writer Dr Johnson said: "An odd thought strikes me—I shall receive no letters in the grave"—but that need no longer be the case. Facebook sites continue already after their owners' deaths, receiving messages and birthday greetings, and many companies exist to continue sending letters after death, as well as to surviving children on their birthdays. Web memorials may be emotionally inadequate, but they may be right for our age and better than a cheap convenience-store bouquet.

Huge chunks of our lives are, of course, already recorded on the Web—not only in the emails and the documents we write, but in our photographs and videos, which we share on sites such as Facebook, Instagram, and YouTube. Companies such as Microsoft, Google, and Apple offer free storage space in return for our life data, so they can access our memories, tastes, and records. But what happens to all this when we die? After years of debriefing its owner's life, a Companion could certainly produce, with current speech technology, a convincing approximation of their voice. It would also have access to a huge store of images, emails, and documents telling its owner's life story. From this, it is not too hard to imagine a Companion continuing after its owner's death to answer questions about their life, and in their own voice.

A Companion that simulated a deceased person might be no more than an updated, computerized form of the goodbye videos now shown at funerals—or on some modern gravestones in Italy, which have, instead of a stone memorial, a small solar-powered video of the deceased activated by a switch. Some may think that a program that assumes the

The Companion's survival of its owner offers the possibility of a kind of immortality—some people are now allowing their Companion avatars to be questioned by relatives at their funerals.

voice and screen image of its deceased owner is an unacceptable form of "immortality." But it seems inevitable that, in the future, the dead will "speak"—so it is worth considering right now what form that conversation should take.

CAN MACHINES BE CONSCIOUS?

The question:

"Will AI artifacts be conscious?"

will not go away, even though no one quite knows what it means, nor how we would know they were conscious. People seem unable to come to a generally agreed conclusion as to whether even dogs are conscious. The question gets tangled up with ethical questions about hunting and food—are lobsters conscious when they go in the pot? The assumption is always that if something were deemed conscious, we could not ill-treat or eat it, and so there would be immediate consequences for how we should treat machines if they achieved conscious status.

Although most of us have an intuitive feel for our own consciousness—even if we cannot describe it well, except to say when we are losing it as we fall asleep or have a presurgical injection—we have no such direct knowledge that *anyone else* is conscious.

As philosophers remind us, other people could all just be zombies who claim to be conscious, just as chatbots have been programmed to do when asked. Stating consciousness proves nothing.

AI scientists and philosophers have tried over decades to map consciousness onto something they *do* understand—such as suggesting machine learning programs may capture the unconscious neural processes of the brain, while logical reasoning captures our conscious planning of actions. But none of that is very convincing, as we seem to do so little logical planning before we act. A more promising approach might come from the claims of American psychologist Julian Jaynes that consciousness has a history and humans did not always consider themselves conscious—and so, consciousness does not automatically come with being *Homo sapiens*. Jaynes' argument, hugely simplified, is that after language was developed—say around 60,000 years ago, although estimates vary widely—humans could talk to themselves in their heads and this puzzled them. That novelty may have been related, Jaynes believes, to the Bible's Old Testament prophets dealing with this new phenomenon by claiming God was talking directly to them. Later, this self-conversation became an essential part of what we now call consciousness, which would imply that only humans have it because only *we* have language.

Computers do not talk to themselves as we do, but we can imagine how they might.

In Chapter 4, when discussing "high-level" languages for programming such as LISP, we saw that the key feature of a language like LISP was that it could be used to express plans or processes that required no specification at all of how "lower-level" languages would translate the LISP code and carry the appropriate actions out in machine code on an actual computer. Similarly, you could say that you plan to run to the street corner, and then do so, but with no idea how your brain and body have translated that intention into nerve and muscle commands—though you know that is what has happened. In the discussion of LISP in Chapter 3, I toyed with the idea that you could say, if you liked the idea, that a language such as English was a very high-level programming language, one where you could say to your automated car, "Take me to Boston," and it would.

Given the above, the possibility that a future AI entity might discuss with itself what it intended to do, weigh up options, and so on, as we do, *but have no idea at all how its machinery would actually carry them out*, is not so implausible at all. It has been claimed that it is just this "sheltering" of our conscious minds from mere detail, such as how we breathe or digest, that allows us to be as effective as we are. That might be the point at which, if an AI machine had such self-discussion and we had evidence of this, we might begin to ask seriously if it were conscious.

TRUTH AND THE ISSUE OF FAKE NEWS IN DIGITAL SPACE

Mark Zuckerberg claims that Facebook will "solve" the problem of Web disinformation, often called "fake news," but this is unlikely, even though a substantial amount of research has gone into detecting the reliability of claims on the Web. That research has not produced any definitive results, and there are conflicting forces at work here. For instance, the early Google PageRank algorithm presented Web pages ordered by the number of other pages that linked to that page and took account of how linked-to those other pages were. The more cited by other pages a Web page was, the more it came out top in a search. This was an original move, which for a while created a notion of a "Google view of truth." It ceased to be truly effective because a way to "spam" the algorithm with spurious links to pop stars was quickly discovered.

This statistical process had, as is often the case, a logical representational equivalent that preceded it: an AI technique called "truth maintenance." The technique assessed how many logically supportive links a statement had from a body of believed truths—that is to say, logical connections that implied it was true, as opposed to the document-to-document links on the Web, hyperlinks as they are usually called, which carry no implications of truth at all, just of mention or repetition of words, to hyperlinks of mention as in PageRank. As is

always the case with logic, this technique was hard to implement on a large scale. However, the idea has not gone away, and if statistical methods fail—as they have so far—to define reliable information on the Web, something of this method will undoubtedly return to favor.

But what has become more important with the Web's aging, is Anglo-Irish satirist Jonathan Swift's old fear that: "Falsehood flies, and truth comes limping after it." More recent research has shown that, on Twitter, false statements are circulated more than true ones. However, Wikipedia, the crowd-sourced encyclopedia, has been a success and can be as reliable as an old-fashioned paper encyclopedia.

The above considerations suggest the picture is mixed, but it may also be that questions about truth and reliability on the Web are often badly posed. A 2018 American survey claimed that the young were better able to sort facts from opinion than the old, but closer examination showed that some of the "facts" in the test were items that many Americans would dispute were true, such as "Obama was born in the US." It is almost certainly true that he was, but, extraordinary though this may seem, many Americans, usually older ones, do not believe this and so were unwilling to classify such a statement as a fact, as opposed to an opinion.

The compilers of the test were naive in not distinguishing facts from what we might call "factual claims," which would cover equally well both the previous (positive) sentence and "Obama was not born in the US."

The underlying point is crucial. It may be that at present large numbers of Western citizens do not agree about what are basic facts (as opposed to opinions, "false facts," or what is now called "fake news"). If true, this is an alarming situation and makes it hard to see how any algorithm from AI could then be expected to decide what is a fact or not, or what is true or false. And since Web settings allow us to see only news that agrees with our viewpoint, such beliefs are not challenged, unless we want them to be. But human beings are very robust. In the old USSR, a population survived even while believing, at the regime's end, that virtually all Soviet media—newspapers, TV, and radio—contained only false news.

The situation is not hopeless. There are sources that can safely be asserted as reliable, from much of Wikipedia to the traditional newspapers. "Reliability" indicators could then be designed along the lines of the PageRank algorithm in terms of an item's links back to such sources, which is how most "news reliability" research currently works.

A related issue is the trustworthiness of Web reviews and recommendations in the world of hotels and restaurants, or indeed any services. The risk now is that many reviews are false, commissioned and paid for by their sponsors, whether positive or negative. There have been style differences detected by NLP techniques between true and fake reviews, but the difference is not reliable. Zuckerberg's hope that Facebook will "solve" the issue of falsehoods on the Web seems both impossible to define accurately and almost inconceivable as a soluble problem.

There is a wider issue—beyond news—of trust in AI artifacts, and this will become central with time. Autonomous cars are today's most pressing example, and many say they will not trust such AI products to keep them safe on the road. Only experience can cure this, as it has with computer-piloted planes and trains, which no longer seem to bother anyone. When the US Army commissioned a fleet of rugged six-legged insect-like robots to carry soldiers' packs over rough terrain, the main problem was not mechanical but that the soldiers did not trust the insects to keep their bags safe.

→
Developed by Ohio State University
during the 1980s, the Adaptive
Suspension Vehicle was intended
as a military all-terrain hydraulic
robot. But poor performance led
to the project being scrapped
in 1990.

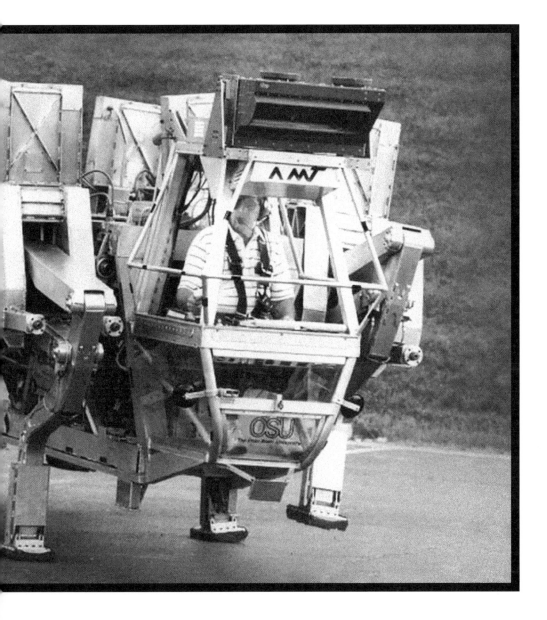

11
SHUTTING UP SHOP

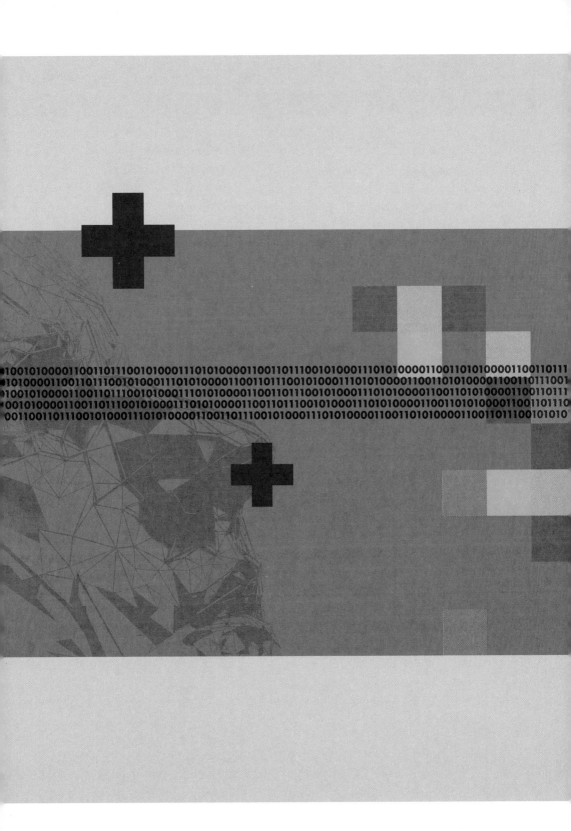

AI has had stunning theoretical and practical successes but is still nowhere near creating what John McCarthy saw as its real goal: to become a form of general human intelligence, something beyond doing specific tasks very well, as it does now, and often better than people. Not all AI workers consider this a possible or even meaningful goal. Nonetheless, along with its half-sibling the Web, AI is utterly disrupting human work, life, and relationships, just as it will transform economies, societies, and warfare on the larger scale. It is also changing the way we see ourselves: what we are, and how and why we function as we do.

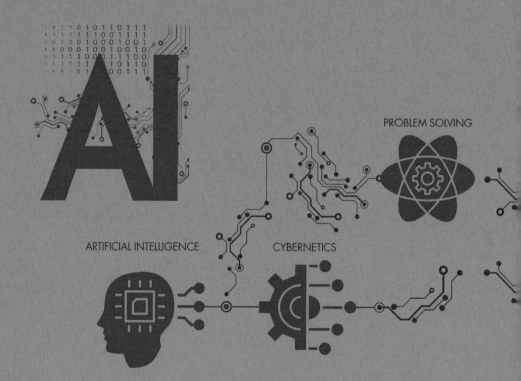

PROBLEM SOLVING

ARTIFICIAL INTELLIGENCE

CYBERNETICS

With the advent of Companion-like agents, we shall accustom ourselves to nonhuman entities being among us, which will make us rethink the metaphysics and ethics (yes, those old concepts) of humanity and of responsibility. Automated cars, already taking to the road in the US and Europe, will make us think some of these thoughts rather earlier.

I emphasized that the history of AI is far longer than many realize, and has had its ups and downs, as well as fashions and freezes on research and funding. It has moved not smoothly but in fits and starts, often overpromising what it can achieve and then being cut back by disillusioned funders. The current fashion is deep learning, which has had great successes but has not yet cracked human language understanding as it did speech recognition and the game of Go. Deep learning cannot be a model of human learning because of the vast amount of data it requires that learner humans plainly do not, and because of its extreme fragility—thrown off course, as we saw in Chapter 6, when a few pixels are added to an image. There already is some swing of the pendulum of fashion in AI thinking, and symbolic representations will come back and be fused with statistical learning in some hybrid method. Then research and development will move forward again. It is important to keep in mind that great advances in the modeling of limited

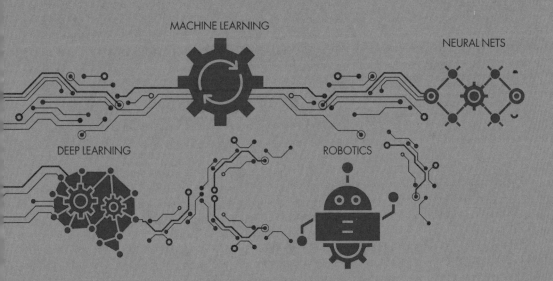

MACHINE LEARNING

NEURAL NETS

DEEP LEARNING

ROBOTICS

skills may not be a step toward general intelligence. As Judea Pearl put it: "The vision systems of the eagle and the snake outperform everything that we can make in the laboratory, but snakes and eagles cannot build an eyeglass or a telescope or a microscope." And just as AI revolutionized psychology in the 1980s by making the idea of symbolic mental representations in a computer central to that discipline, so it is now revolutionizing sociology, with models of networks of individuals and groups and how influence spreads among them—work pioneered by American computer scientist Jon Kleinberg at Cornell University.

This movement is influencing the behavior of media giants such as Facebook, as well as public attitudes to them, and the future of the Web is unclear. For example, political and financial constraints on these companies in the Western world might conceivably make the Chinese system Baidu our main platform at some future point. Our governments see the technical giants as cash cows for tax purposes but have no idea how to moderate the changes they are creating: in social disinhibition and emotion, in political influence, and in the mental life of children. Stephen Pinker has argued that despite all this, the world is drawing together and behaving better overall, and we can only hope he is right.

What is clear is that the data that a single human life now creates and accesses is becoming too great to be managed by that individual, and some assistance, maybe Companion-like, will be needed by most people. It is also clear that this control of our data and our identity will be a crucial political question of coming decades. Politicians have been slow to understand what is happening and to face up to any kind of responsibility to control the results of radical automation. Many still want to tax and deter it—which is precisely the wrong answer—and are slow to face up to the reality of general surveillance of the population. They continue to deny, in countries such as the US and UK, that their agencies access the communications of all citizens at will and without control, as they surely do. Google has admitted it "controls the majority of online conversations." AI has created all these problems and will have to be part of the solution.

CONVERSATIONS WITH AI

The thoughts of Alan Turing are as good a place as any to date the start of AI and maybe also to end this book. The bar on real relationships with machines is quite low, as I noted, with primitive pre-Companions such as Tamagotchi, but Turing also wanted in 1950 to set the bar high, and wrote that a computer should be able to conduct the following conversation, one which is still well beyond our current technical methods:

INTERROGATOR:

In the first line of your sonnet, which reads "Shall I compare thee to a summer's day," would not "a spring day" do as well or better?

COMPUTER:
It wouldn't scan.

INTERROGATOR:

How about "a winter's day?" That would scan all right.

COMPUTER:

Yes, but nobody wants to be compared to a winter's day.

INTERROGATOR:

Would you say Mr Pickwick reminded you of Christmas?

COMPUTER: In a way.

INTERROGATOR:

Yet Christmas is a winter's day, and I do not think Mr Pickwick would mind the comparison.

COMPUTER:

I don't think you're serious. By a winter's day one means a typical winter's day, rather than a special one like Christmas.

INDEX

INDEX

ACKNOWLEDGMENTS

IMAGE CREDITS

I owe a debt of gratitude to Peet Morris who encouraged me to take on this book, as well as to Ken Ford for his continuing support over the years as Director of the Florida Institute of Human and Machine Cognition. I owe a great deal to the team at Icon (Duncan Heath, Brian Clegg, and Rob Sharman) for their criticisms and suggestions. I also owe thanks to the many people who have made comments and criticisms on drafts of the book: Peet Morris, Patrick Hanks, Arthur Thomas, Robert Hoffman, Greg Grefenstette, Sergei Nirenburg, David Levy, John Tait, Tom Wachtel, Alexiei Dingli, Nick Ostler, Christine Madsen, Tomek Strzalkowski, Richard Weyhrauch, Derek Partridge, Angelo Dalli, and Ken Lovesy. The errors, of course, are all my own.

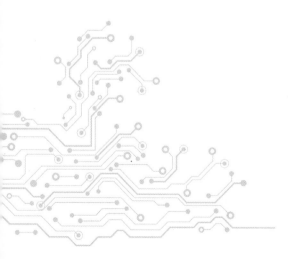